# True Adventure Stories

## 探險家的故事

商務印書館

U0108891

Name of Book: True Adventure Stories
Editor: Daniela Penzavalle
Design and art direction: Nadia Maestri
Computer graphics: Simona Corniola
Picture research: Laura Lagomarsino
Edition: ©2008 Black Cat Publishing,
          an imprint of Cideb Editrice, Genoa, Canterbury

系　列　名：Black Cat 優質英語階梯閱讀 · Level 3
書　　　名：探險家的故事
責任編輯：畢　琦
封面設計：張　毅
出　　　版：商務印書館（香港）有限公司
　　　　　　香港筲箕灣耀興道 3 號東滙廣場 8 樓
　　　　　　http://www.commercialpress.com.hk
發　　　行：香港聯合書刊物流有限公司
　　　　　　香港新界大埔汀麗路 36 號中華商務印刷大廈 3 字樓
印　　　刷：中華商務彩色印刷有限公司
　　　　　　香港新界大埔汀麗路 36 號中華商務印刷大廈
版　　　次：2009 年 1 月第 1 版第 1 次印刷
　　　　　　© 2009 商務印書館（香港）有限公司
　　　　　　ISBN 978 962 07 1855 7
　　　　　　Printed in Hong Kong

# 出版說明 ─────────────

本館一向倡導優質閱讀，近年來連續推出了以 "Q" 為標識的 "Quality English Learning 優質英語學習" 系列，其中《讀名著學英語》叢書，更是香港書展入選好書，讀者反響令人鼓舞。推動社會閱讀風氣，推動英語經典閱讀，藉閱讀拓廣世界視野，提高英語水平，已經成為一種潮流。

然良好閱讀習慣的養成非一日之功，大多數初中級程度的讀者，常視直接閱讀厚重的原著為畏途。如何給年輕的讀者提供切實的指引和幫助，如何既提供優質的學習素材，又提供名師的教學方法，是當下社會關注的重要問題。針對這種情況，本館特別延請香港名校名師，根據多年豐富的教學經驗，精選海外適合初中級英語程度讀者的優質經典讀物，有系統地出版了這套叢書，名為《Black Cat 優質英語階梯閱讀》。

《Black Cat 優質英語階梯閱讀》體現了香港名校名師堅持經典學習的教學理念，以及多年行之有效的學習方法。既有經過改寫和縮寫的經典名著，又有富創意的現代作品；既有精心設計的聽、說、讀、寫綜合練習，又有豐富的歷史文化知識；既有彩色插圖、繪圖和照片，又有英美專業演員朗讀作品的 CD。適合口味不同的讀者享受閱讀之樂，欣賞經典之美。

《Black Cat 優質英語階梯閱讀》由淺入深，逐階提升，好像參與一個尋寶遊戲，入門並不難，但要真正尋得寶藏，需要投入，更需要堅持。只有置身其中的人，才能體味純正英語的魅力，領略得到真寶的快樂。當英語閱讀成為自己生活的一部分，英語水平的提高自然水到渠成。

商務印書館（香港）有限公司
編輯部

# 使用說明 _____

**1** 應該怎樣選書？

### 按閱讀興趣選書

《Black Cat 優質英語階梯閱讀》精選世界經典作品，也包括富於創意的現代作品；既有膾炙人口的小說、戲劇，又有非小說類的文化知識讀物，品種豐富，內容多樣，適合口味不同的讀者挑選自己感興趣的書，享受閱讀的樂趣。

### 按英語程度選書

《Black Cat 優質英語階梯閱讀》現設Level 1 至 Level 6，由淺入深，涵蓋初、中級英語程度。讀物分級採用了國際上通用的劃分標準，主要以詞彙（vocabulary）和結構（structures）劃分。

Level 1 至 Level 3 出現的詞彙較淺顯，相對深的核心詞彙均配上中文解釋，節省讀者查找詞典的時間，以專心理解正文內容。在註釋的幫助下，讀者若能流暢地閱讀正文內容，就不用擔心這本書程度過深。

Level 1 至 Level 3出現的動詞時態形式和句子結構比較簡單。動詞時態形式以簡單現在式（present simple）、現在進行式（present continuous）、簡單過去式（past simple）為主，句子結構大部分是簡單句（simple sentences）。此外，還包括比較級和最高級（comparative and superlative forms）、可數和不可數名詞（countable and uncountable nouns）以及冠詞（articles）等語法知識點。

Level 4至 Level 6 出現的動詞時態形式，以現在完成式（present perfect）、現在完成進行式（present perfect continuous）、過去完成進行式（past perfect continuous）為主，句子結構大部分是複合句（compound sentences）、條件從句（1st and 2nd conditional sentences）等。此外，還包括情態動詞（modal verbs）、被動式（passive forms）、動名詞

（gerunds）、短語動詞（phrasal verbs）等語法知識點。

根據上述的語法範圍，讀者可按自己實際的英語水平，如詞彙量、語法知識、理解能力、閱讀能力等自主選擇，不再受制於學校年級劃分或學歷高低的約束，完全根據個人需要選擇合適的讀物。

## ② 怎樣提高閱讀效果？

閱讀的方法主要有兩種：一是泛讀，二是精讀。兩者各有功能，適當地結合使用，相輔相成，有事半功倍之效。

泛讀，指閱讀大量適合自己程度（可稍淺，但不能過深）、不同內容、風格、體裁的讀物，但求明白內容大意，不用花費太多時間鑽研細節，主要作用是多接觸英語，減輕對它的生疏感，鞏固以前所學過的英語，讓腦子在潛意識中吸收詞彙用法、語法結構等。

精讀，指小心認真地閱讀內容精彩、組織有條理、遣詞造句又正確的作品，着重點在於理解"準確"及"深入"，欣賞其精彩獨到之處。精讀時，可充分利用書中精心設計的練習，學習掌握有用的英語詞彙和語法知識。精讀後，可再花十分鐘朗讀其中一小段有趣的文字，邊唸邊細心領會文字的結構和意思。

《Black Cat 優質英語階梯閱讀》中的作品均值得精讀，如時間有限，不妨嘗試每兩個星期泛讀一本，輔以每星期挑選書中一章精彩的文字精讀。要學好英語，持之以恆地泛讀和精讀英文是最有效的方法。

## ③ 本系列的練習與測試有何功能？

《Black Cat 優質英語階梯閱讀》特別注重練習的設計，為讀者考慮周到，切合實用需求，學習功能強。每章後均配有訓練聽、説、讀、寫四項技能的練習，分量、難度恰到好處。

聽力練習分兩類，一是重聽故事回答問題，二是聆聽主角對話、書信朗讀、或模擬記者訪問後寫出答案，旨在以生活化的練習形式逐步提高聽力。每本書均配有 CD 提供作品朗讀，朗讀者都是專業演員，英國作品由英國演員錄音，美國作品由美國演員錄音，務求增加聆聽的真實感和感染力。多聆聽英式和美式英語兩種發音，可讓讀者熟悉二者的差異，逐漸培養分辨英美發音的能力，提高聆聽理解的準確度。此外，模仿錄音朗讀故事或模仿主人翁在戲劇中的對白，都是訓練口語能力的好方法。

閱讀理解練習形式多樣化，有縱橫字謎、配對、填空、字句重組等等，注重訓練讀者的理解、推敲和聯想等多種閱讀技能。

寫作練習尤具新意，教讀者使用網式圖示（spidergrams）記錄重點，採用問答、書信、電報、記者採訪等多樣化形式，鼓勵讀者動手寫作。

書後更設有升級測試（Exit Test）及答案，供讀者檢查學習效果。充分利用書中的練習和測試，可全面提升聽、說、讀、寫四項技能。

## ④ 本系列還能提供甚麼幫助？

《Black Cat 優質英語階梯閱讀》提倡豐富多元的現代閱讀，巧用書中提供的資訊，有助於提升英語理解力，擴闊視野。

每本書都設有專章介紹相關的歷史文化知識，經典名著更附有作者生平、社會背景等資訊。書內富有表現力的彩色插圖、繪圖和照片，使閱讀充滿趣味，部分加上如何解讀古典名畫的指導，增長見識。有的書還提供一些與主題相關的網址，比如關於不同國家的節慶源流的網址，讓讀者多利用網上資源增進知識。

# Contents

The text is recorded in full.　故事錄音

🎧 END　These symbols indicate the beginning and end of the passages
linked to the listening activities.　聽力練習開始和結束的標記

# The Spirit of Adventure

In the past thirty years the population of our planet has grown by 62%. Today it is six billion. In 2055 it will be over eight billion. At the same time, modern technology has made our world smaller. These two facts mean that there are not many places left where we can find real adventure. Those people who dream of adventure go on adventure holidays to look for it. But this fantasy adventure [1] is very different from the real thing.

So what is the real thing? It is a challenge against the unknown with only a small chance of success. It is a struggle with great hardship [2], danger and even death.

A very good example is given by Alexander Laing (1793-1826), the first European to reach Timbuktu in 1826. During his journey he was attacked by robbers, nearly died of hunger and thirst and was very ill with malaria [3]. He managed to arrive in Timbuktu but was then killed by Arabs on his return journey. Another adventurer, René Caillé (1799-1839), made a dangerous journey across the Sahara in temperatures of 70° C disguised [4] in Muslim clothes and managed to see Timbuktu and return alive. Many others who had the spirit of adventure paid with their lives, like Captain James Cook (1728-79), who was murdered on Hawaii during his third exploratory voyage in the Pacific.

Others died of hunger and disease, like Robert O'Hara Burke (1821-61) and William John Wills (1834-61) when they tried to cross Australia from Melbourne in the south to the Gulf of Carpentaria in the north, at a time when most of the country was completely unknown to European settlers.

1. **fantasy adventure**：假想探險。
2. **hardship**：艱苦境地。
3. **malaria**：瘧疾。
4. **disguised**：喬裝假扮。

Some just disappeared, like Colonel Percy Fawcett (1867-1925) who went on an expedition in 1925 to find what he believed to be an ancient lost city in the jungles of Brazil. He was never seen again.

Why did all these adventurers decide to face danger and death? There are many different reasons: curiosity [1], ambition [2], the need to be famous, or to follow their romantic dreams. But all of them were trying to escape from ordinary life for the excitement of adventure.

In these stories you'll read of people who challenged the unknown and succeeded in finding the real thing, like

**Sir Ernest Shackleton** photographed in 1908.

Ernest Shackleton (1874-1922) and his ordeal [3] in Antarctica, which a polar traveller with modern communications, food and equipment would not have to experience today [4].

The pioneer airwoman Amy Johnson (1903-41) set off alone in a

1. **curiosity** : 好奇。
2. **ambition** : 野心。
3. **ordeal** : 嚴酷的考驗；此處指苦旅。
4. **which a polar traveller ... today** : 這是如今擁有現代通訊器材、食物和設備的極地旅行家不必再經歷的。

**Amy Johnson** photographed in May 1930 at Croydon Airport, England.

single engine airplane from Croydon, London, on 5 May 1930, and landed in Darwin on 24 May, an epic [1] flight of 11,000 miles. She was the first woman to fly alone to Australia.

The great explorer David Livingstone (1813-73) was almost obsessed [2] with finding the source of the Nile.

An English farmer, Dougal Robertson, simply wanted to leave his ordinary life in England and set off with his wife and children to sail round the world in 1971. Their adventure became a battle for survival. Luckily, they won. But like all those who have had the experience of real adventure, Dougal felt sad when it was over.

1. **epic** : 壯麗的。　　　　2. **be obsessed** : 着迷（多用作被動結構）。

### ➊ Comprehension check
**Say if the following sentences are true (T) or false (F).**

|  |  | T | F |
|---|---|---|---|
| 1 | In the next fifty years the world's population will grow by eight billion. | ☐ | ☐ |
| 2 | Fantasy adventure is not the same as the real thing. | ☐ | ☐ |
| 3 | All of the adventurers were either killed, died of hunger and disease, or disappeared. | ☐ | ☐ |
| 4 | Most of the adventurers were looking for excitement. | ☐ | ☐ |
| 5 | The adventures of Shackleton could not happen today. | ☐ | ☐ |
| 6 | The final sentence suggests that when a real adventure is over, you want to forget it. | ☐ | ☐ |

# Lost at Sea

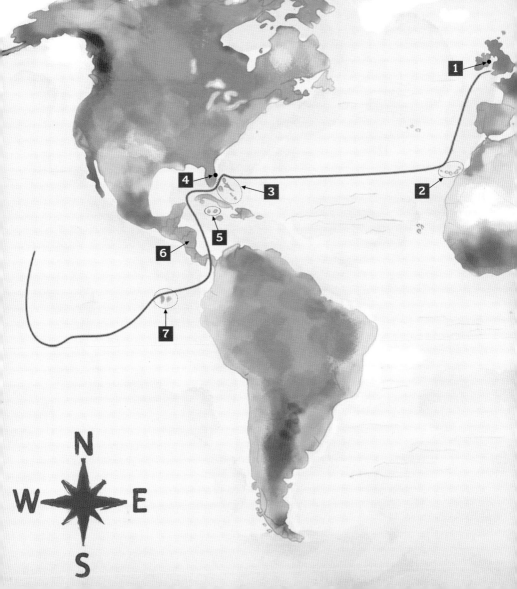

# Before you read

### ❶ Reading maps

Look at the map on page 11. Can you say where these places are?

> Falmouth    the Canary Islands    the Bahamas    Miami
> Jamaica    Panama    the Galapagos Islands

### ❷ The characters

Look at the picture of the Robertson family in 1968. Match the descriptions (1-6) to the people (A-F).

1 ☐  Neil is nine years old and he is Sandy's twin. His hair is short and blond.
2 ☐  Anne is sixteen. She's got long, light brown hair.
3 ☐  Lyn is thirty-eight years old. Her hair is long and dark.
4 ☐  Douglas is fifteen. His hair is black.
5 ☐  Sandy is nine years old and Neil's twin. He's got long blond hair.
6 ☐  Dougal is forty-four. He's got brown hair and a beard.

## PART ONE

# Abandon Ship!

**D**ougal Robertson had a small farm in the north of England. After fifteen years as a farmer, 44-year-old Dougal was not happy with his life. He worked very hard, but he could not make any money. He often said to himself, 'Farming is a very difficult life. I've had only two weeks' holiday in fifteen years! All my hopes for a good life have gone.'

One Sunday morning in 1968 Dougal heard some news on the radio about a round-the-world yacht [1] race. His family listened while he told them about the race. He tried to describe what it was like sailing alone in the wide, empty ocean. Then he and his wife Lyn told the children about their sailing adventures in Hong Kong before they had the farm. Suddenly Neil, one of their nine-year-old twins, shouted, 'Daddy's a sailor! Why can't we go round the world?'

Sandy, the other twin, said, 'That's a great idea! Let's buy a boat and go round the world!'

1. **yacht** : 遊艇。

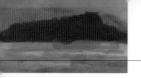
Everybody started laughing, but Dougal became quiet. He looked at his daughter Anne, 16, and his son Douglas, who was 15. They had never been far from home and knew nothing about the big world.

'Why not?' Dougal thought. 'Let's have a great adventure and sail round the world! The children will see some wonderful things and meet lots of different people. They will learn interesting things about life and the world. I was a sailor for twelve years and I've got a Master Mariner's certificate [1]. Lyn is a nurse. We can speak a few foreign languages, though not very well! After fifteen years on the farm we're all in good physical condition. Yes! We'll do it!'

So during the next two years Dougal and Lyn sold the farm and made preparations for a voyage round the world. Dougal bought a small schooner [2] called *Lucette*, which was 12 metres long. Although fifty years old, after some repairs she was in perfect condition. There was also a life raft [3] on board that inflated automatically.

In January 1971, when the Robertsons left Falmouth, a port in Cornwall, everybody was excited. They began their voyage across the Atlantic from the Canary Islands. Then they visited the West Indies and the Bahamas, where Anne fell in love with a young man from Canada. When they stopped in Miami for a while, Dougal bought a fibreglass dinghy [4].

'Who knows? Maybe our lives will depend on it one day,' Lyn said, laughing.

---

1. **certificate** : 證書。

2. **schooner** : 雙桅船。

3. **life raft** :  救生筏。

4. **fibreglass dinghy** : 玻璃鋼（一種很輕的材料）舢舨。

In February 1972 Anne, who was nearly twenty years old, decided to stay in the Bahamas with her Canadian boyfriend. The rest of the family sailed on to Jamaica, and then south to Panama. There, a 22-year-old student called Robin Williams, who was travelling around the world, asked the Robertsons to take him to New Zealand. One morning in May 1972 they sailed into the Pacific Ocean towards the Galapagos Islands.

On 13 June the *Lucette* left the Galapagos Islands to begin her voyage to the Marquesas Islands, nearly 5,000 kilometres to the west. Now the Robertsons were alone in the largest ocean in the world. At 9.55 am on 15 June Dougal was in his cabin when there was a loud crash. Something had hit the floor. As Dougal fell against the bed, he heard Douglas shout 'Whales!' When he looked under the floor, he saw sea water coming through a big hole in the bottom of the boat. He tried to stop it, but it was impossible. The *Lucette* was sinking fast.

'Abandon ship!' he shouted.

In a terrible panic [1] everybody put on their lifejackets [2], pushed the dinghy into the sea and threw into it anything they could find. There was also the life raft, which they all swam to. One minute later the *Lucette* sank. It was 10 o'clock.

Douglas said, 'Killer whales [3]. I think three of them attacked us.'

Everybody looked shocked and frightened. The twins were crying.

'We must get these boys home,' said Lyn quietly.

'Of course,' Dougal replied. 'We'll make it!'

---

1.  **panic** : 恐慌。
2.  **lifejackets** : 救生衣。
3.  **Killer whales** : 虎鯨。

But when he thought about their situation, he was not so certain. They were more than 300 kilometres from the Galapagos Islands. The Marquesas Islands were 4,600 kilometres away. He had no compass [1] and no maps for navigation. There was a survival kit [2] with some food, water and fishing things. They had a knife, biscuits, onions, some sweets and fruit.

'We can live for maybe ten days on the food and water,' he thought. 'We're a very long way from the international shipping routes. There may be no rain in this region for six months. Without water we will die.'

'What are our chances of survival?' Lyn asked him. 'Tell us the truth.'

But Dougal could not tell them what he thought. Then he suddenly knew that there was only one thing to do. He said they must go north to the Doldrums, about 600 kilometres away. That was their only chance of finding rain water and the only hope of meeting a ship. Now everybody began to feel better. The twins stopped crying and looked at the sweets.

During the first week there were many problems. They were always hungry and thirsty. The waves were more than four metres high and Robin and Neil were sea-sick. It was very uncomfortable in the small raft. Big fish called dorado swam against the bottom all day and night and sometimes bit them. Everybody slept badly. There were small holes in the raft where the air escaped [3], so they had to blow it up [4] often. Also, they had to throw out the sea water which was coming through the floor.

---

1. **compass :** 指南針。
2. **survival kit :** 救生工具包。
3. **the air escaped :** 漏氣。
4. **blow it up :** 充氣。

On the third day a flying fish fell into the dinghy behind the raft. They ate it for breakfast. During the next days there were more flying fish, and on the sixth day Dougal found a big dorado in the dinghy. They ate it with lemon juice, a small piece of onion, the last orange and a drop of water each.

After a week their clothes were in a bad condition. Their skin was sore [1] from the sea water. They were tired from the nights without sleep and their bodies ached. Then one day it rained. They collected it and drank a lot. The rain also cleaned their faces and hair. Water! They all felt much better.

Later that day Douglas shouted, 'A ship! A ship!'

It was more than four kilometres away. Dougal lit a rocket flare [2] and then a second one. But the ship did not see them. When it disappeared, they all looked silently at the horizon.

From that moment Dougal changed. He said to the others, 'If a ship can't rescue us, we must do it alone. The new word is survival, not rescue or help. We will use our intelligence and beat this terrible sea.'

'What's the password [3] for today?' Lyn said, and everyone shouted back 'Survival!'

In the evening they caught a turtle and killed it. Next morning Dougal cut it up and found a hundred yellow eggs inside. They had turtle meat and eggs for breakfast. For the first time their stomachs were full.

1. **sore**：疼痛。
2. **rocket flare**：照明彈。通常用作船隻信號。
3. **password**：此處指口號。

# The text and **beyond**

**1 Comprehension check**

Match the first half of each sentence (A-J) with the second half (1-10) to make a summary of Part One.

A ☐ Dougal Robertson was not happy with his life as a farmer because

B ☐ One Sunday, after Dougal's children heard some news about a yacht race

C ☐ Two years later the Robertsons began their voyage

D ☐ On 15 June 1972, two days after they left the Galapagos Islands,

E ☐ Dougal knew the situation was extremely dangerous,

F ☐ Then he said they must go 600 kilometres north to the Doldrums,

G ☐ At first, they all suffered a lot

H ☐ After a week their clothes and bodies were in a bad condition,

I ☐ When a ship did not see them, Dougal said

J ☐ Later, they caught a turtle which was full of eggs,

1 but he could not tell his family the truth.

2 that they had to survive alone and beat the sea.

3 they said that they wanted to sail round the world.

4 but luckily they ate some fish and drank when it rained.

5 and sailed across the Atlantic in a small schooner called *Lucette*.

6 where they might find water and meet a ship.

7 so for the first time they were not hungry.

8 after working hard for fifteen years he had not made any money.

9 three killer whales attacked the boat, which sank.

10 because they were always hungry and thirsty and could not sleep.

**2** **What happened there?**
Can you remember what happened in these places? Write a sentence for each place, then check your answers in the text. There is an example at the beginning (0).

0   Hong Kong: .Dougal and his wife Lyn had some sailing adventures. ....
1   Falmouth, Cornwall: ...................................................................................
2   The Canary Islands: ..............................................................................
3   The Bahamas: ........................................................................................
4   Miami: ...................................................................................................
5   Panama: .................................................................................................
6   The Galapagos Islands: ........................................................................

**PET** **3** **Sentence transformation**
Here are some sentences from Part One. Complete the second sentence so that it means the same as the first. Use no more than three words. There is an example at the beginning (0).

0   'Why don't we go round the world!' said Neil.
    'Let's go............................................... round the world!' said Neil.
1   'I think three killer whales attacked us,' Douglas said.
    'I think ...................................... by three killer whales,' Douglas said.
2   Dougal bought a small schooner called *Lucette*, which was 12 metres long.
    *Lucette*, the small schooner ............................., was 12 metres long.
3   Everybody slept badly in the small raft.
    Nobody ....................................................... in the small raft.
4   'What are our chances of survival?' Lyn asked Dougal.
    Lyn asked Dougal what their chances ............................................. .
5   'Let's buy a boat and go round the world!' said Sandy.
    'Why ........................................ buy a boat and go round the world!' said Sandy.
6   'Without water we will die,' Dougal thought.
    '............................................ find water we will die,' Dougal thought.

# *The Enchanted Islands* [1]

A group of thirteen main islands and six smaller ones, the Galapagos Islands are located in the Pacific Ocean, 1,100 kilometres west of Ecuador in South America. Since 1832 they have been a province of Ecuador and they belong to that country's national park system. The capital is San Cristobal Island, with a population of 2,000, and four other islands are also inhabited [2].

The Galapagos Islands are famous all over the world for many reasons. They were once known as 'the Enchanted Islands' because navigation was very difficult in the sea around them. They are located at a place where volcanos have been appearing from under

1. **the Enchanted Islands** : 魔島。          2. **be inhabited** : 有人居住。

**From top left:** flamingoes; seal; booby; sea lions.

the sea for millions of years. The oldest island – Española – is about three and a half million years old and during this time it has moved slowly south. The youngest islands are Isabel and Fernandina, which are still forming.

The Galapagos have an enormous variety of [1] plant and animal life and many species are unique [2] to each island. For example, a bird called the 'vampire finch' is found only on Wolf Island. It has adapted [3] to the island's environment by drinking the blood of another bird called the 'booby'.

Similarly, there are eleven different types of the famous giant tortoises, which gave the islands their name. Each species of tortoise has adapted to its island, so their shells now have different shapes.

1. **an enormous variety of**：大量各種各樣的。
2. **unique**：獨一無二的。      3.  **adapted**：適應。

**From top left:** albatross; frigate bird; pelican; marine iguana.

There are also marine iguanas, seals, sea lions, green turtles, pink flamingos, tropical penguins, pelicans, whales, dolphins and thousands of species of birds.

In 1835, when a young English naturalist visited the Galapagos during a historic voyage on the Beagle, he noticed small variations [1] in the finches of several islands. His observations were the inspiration for his theory of evolution, which he presented in his book *The Origin of Species* (1859). His name was Charles Darwin.

Discovered by the Spanish in 1535, for many years the islands were used as a base by English pirates to attack Spanish ships. After the pirates came whalers. Since the 18th century there has been a wooden barrel [2] at Post Office Bay on Floreana: it was first used by whalers for

1.  **variations**：變化。        2.  **barrel**：平底平頂的桶。

**Lonesome George.** 'Lonesome' is the American word for the British 'lonely'.

posting letters. These whalers killed or captured thousands of giant tortoises for their fat. Seals, too, became almost extinct [1].

The island of Pinta is home to the world's rarest creature: one giant Pinta tortoise called Lonesome George. He is very old, and when he dies the species will be extinct.

Today the greatest danger to the environment of the islands is the plants and animals introduced by visitors over the centuries. Goats, pigs, cows, dogs, cats, rats, sheep and horses have gradually destroyed the habitats [2] of native species. On Santiago island pigs are responsible for the extinction of the land iguanas which were

---

1. **extinct** : 絕種。　　　　　2. **habitats** : 棲息地。

numerous in Darwin's time. There is tourism, of course, and also the human population explosion [1].

In 1959 the Galapagos Islands had approximately [2] 1,000-2,000 inhabitants. Today the population is 30,000. Nevertheless, the islands are still full of wildlife.

Fortunately, 70,000 square kilometres of ocean around the islands became a marine reserve in 1986 and in 1990 the region was made a whale sanctuary [3]. Since 1978 the Galapagos have been a UNESCO World Heritage Site.

1. **the human population explosion** : 人口爆炸。
2. **approximately** : 大約。
3. **sanctuary** : (動物) 保護區。

### 1 Comprehension check

**Say whether the following statements are true (T) or false (F). Then correct the false ones.**

|  |  | T | F |
|---|---|---|---|
| 1 | There are thirteen islands, five of them inhabited. | ☐ | ☐ |
| 2 | The Galapagos are volcanic islands. | ☐ | ☐ |
| 3 | All the islands are over three million years old. | ☐ | ☐ |
| 4 | The vampire finch lives only on one island. | ☐ | ☐ |
| 5 | The islands are named after a species of giant shells. | ☐ | ☐ |
| 6 | *The Origin of Species* is a journal of Darwin's visit to the Galapagos. | ☐ | ☐ |
| 7 | Whalers helped to preserve the islands' wildlife population. | ☐ | ☐ |
| 8 | During the last fifty years the human population of the islands has increased more than ten times. | ☐ | ☐ |
| 9 | The islands' sea mammals include seals, sea lions, turtles and whales. | ☐ | ☐ |
| 10 | In the last thirty years nothing has been done to help the region's wildlife. | ☐ | ☐ |

# Before you read

### 1 Prediction

**Before you continue, try to guess what is going to happen to the Robertsons.**

1 An aeroplane will see them and they will all be rescued.
2 They will all survive and a ship will rescue them.
3 One or two of them will not survive.

### 2 Listening

**Listen to the beginning of Part Two. For questions 1-5 choose the correct answer — A, B or C.**

1 The Robertsons poured water over their bodies
   A ☐ because they were thirsty.
   B ☐ to cool themselves.
   C ☐ to wash themselves.

2 Dougal caught a shark with
   A ☐ a spear.
   B ☐ his hands.
   C ☐ a paddle.

3 To pass the time the Robertsons
   A ☐ talked and played games.
   B ☐ repaired the raft.
   C ☐ moved more slowly.

4 Dougal had to swim fast because
   A ☐ some sharks were following him.
   B ☐ the dinghy was fifty metres away.
   C ☐ the rope had broken.

5 The Robertsons left the raft because
   A ☐ it was sinking.
   B ☐ it couldn't hold them and their provisions.
   C ☐ the dorado bit them.

## PART TWO

# The Battle to Survive

In the second week the weather was sunny. They sat under the raft's canopy [1] and poured water over their hot bodies. But they had to continue throwing out water and blowing up the raft. When Dougal threw out some pieces of turtle, four big sharks arrived. He hit one with a paddle and it swam away. Later, he made a fishing spear and tried to catch a shark, but the spear did not work well. Then he tried with his hands. Next moment, to his surprise, he had a 1.5 metre shark in his arms — and let it go!

After two weeks everyone was weaker and moved more slowly. They were all much thinner and had inflamed [2] skin from the sea water. The raft was now in a worse condition. They had to repair new holes every day. Dougal asked himself, 'How much longer have we got before it becomes impossible to live in?' To pass the time they played games like Twenty Questions and

1. **canopy** : 一種帳篷。
2. **inflamed** : 紅腫的。

I Spy. And their conversations [1] were always about food and drink: cold milk, ice cream, fresh fruit...

When they caught another turtle, they drank the blood. On day fifteen there was heavy rain and they drank and drank. But then they noticed that the dinghy was not behind them. The rope that tied it to the raft had broken. It was about fifty metres away. As Dougal swam towards it, Douglas shouted, 'Sharks!' With two sharks following him Dougal swam for his life. He reached the dinghy safely and paddled [2] it back to the raft.

Dougal knew they had to move to the little dinghy soon. He asked himself if it could hold six people and all the supplies without sinking. But he knew there was no other choice. So on day seventeen he decided it was time to leave the raft. They put everything they needed into the dinghy. Now, as they sadly watched the raft sink, they had only the sky for a roof. But the dinghy seemed stable [3], and it was dry. It was also hard, so the dorado could not bite them now!

ᴇɴᴅ

4 July was Lyn's birthday. They caught another turtle, had a meal and sang songs. They were used to living from the sea now, and Dougal felt certain that they could 'get the boys home.' Their clothes were falling off their bodies, the nights were cold, but they could survive in the dinghy. Two days later they caught a turtle with a hundred eggs in it. Robin drank three cups of blood like a vampire. For the second time their stomachs were full. But in the afternoon the wind blew stronger and the waves became very high. They had to throw out a lot of water from the bottom

1. **conversations** : 談話。
2. **paddled** : 划船。
3. **stable** : 堅固。

of the dinghy. Then at 10pm came a storm with heavy rain, thunder and lightning.

They passed a very cold, wet night. While Sandy cried and Lyn prayed, Dougal tried to steer [1] the dinghy. He was freezing, so Lyn told Robin to rub [2] him. But Dougal could not feel anything. Then Douglas shouted, 'Sing to keep warm!' So they started singing.

The next night the storm returned. As they threw out water for hours, they sang and shivered [3] and Lyn rubbed their bodies. When day came, the rain stopped and they were all asleep. That afternoon the sun came out; finally they could dry their things. Everybody knew that they had been very close to death.

Now the weather was hot and the sea was calm. But as the hot weather continued, Dougal and Lyn began to quarrel. There were angry words between them about their marriage and their life on the farm. When Sandy shouted at them to stop, Dougal said, 'Lyn, be quiet or I'll leave you and go to sea!' Then everybody laughed.

There were only a few litres of water left now. The hot sun shone down every day. Their tongues felt big and like leather in their mouths. They were so thirsty that when Dougal caught another three dorado, they drank the fluid from the eyes and bones. When another turtle appeared, Douglas lost it. His father shouted angrily and hit him. Lyn cried and called Dougal a bully [4]. On day thirty-five Robin also lost a turtle. Furious, Dougal hit

1. **steer** : 駕駛。
2. **rub** : 摩擦取暖。
3. **shivered** : 顫抖。
4. **bully** : 惡霸。

him on the face. Taking a paddle, he shouted, 'I'll hit you with this if you do it again!' In the stress of the situation Dougal was becoming desperate [1]. He knew how much their lives depended on care and attention.

Day thirty-six was cold after a night of wind and rain. After more than a month at sea they all looked like primitive [2] people. Douglas had only a shirt, Dougal's clothes were covered in blood and turtle fat, and he and Robin had long beards. Everyone was dirty and had cuts and scratches [3]. But they were still alive and their position was near the shipping route between Panama and Hawaii. They talked for hours about their favourite food. On the thirty-seventh day they caught another turtle. Dougal took three hours to cut it up. He was weaker and needed to rest often.

On day thirty-eight Dougal caught a big dorado. Lyn washed the twins and made them exercise their muscles [4]. After the evening meal Dougal suddenly said, 'A ship. There's a ship and it's coming towards us!' Nobody could believe it. Dougal told everyone to sit still and not to sink the dinghy. Carefully he stood up and asked for a flare.

'Oh God!' said Lyn. 'Please let them see us!'

The ship was about one and a half kilometres away. Everybody was very excited and nervous. Then Dougal saw two large sharks not far away. 'Be careful, we mustn't sink now!' he said. 'I'm going to light the flare.' It lit up the evening with a red light.

'Quick, give me another!' he said. 'I think she's changing direction.'

1. **desperate** : 絕望的。
2. **primitive** : 原始的。
3. **scratches** : 抓痕。
4. **muscles** : 肌肉。

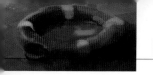

But the ship made no sign that she had seen them. Dougal was so nervous he felt nausea [1] in his stomach. When the second flare did not work, he shouted, 'The torch! We'll signal to them.' At that moment he saw that the ship was really coming towards them, and he fell down.

'It's over!' he said quietly.

Everybody was crying with happiness. With tears in his eyes Dougal put his arms round Lyn. 'We'll get these boys home after all,' he said.

The sailors of the *Tokamaru*, a Japanese fishing ship, helped them to climb on board. They had to lie on the deck because they could not walk. And they smelt bad. So later they washed with hot water and soap and dressed in clean, dry clothes — a wonderful luxury! Then they had big plates of bread and butter and lots of coffee. Everything was so strange they could not sleep. They needed time to return to normal life.

During the next four days, as the *Tokamaru* sailed to Panama, the Robertsons ate a lot of delicious food. But after every big meal they always felt hungry. They also exercised their bodies and learnt to walk again. They had travelled more than 1,000 kilometres lost and alone on the Pacific Ocean. It had been a terrible fight against the sea, but they had won. Now they were free. But for a long time afterwards Dougal felt a strange nostalgia [2] for the lonely, dangerous life on the sea.

1. **nausea** : 噁心。
2. **nostalgia** : 懷舊情緒。

# The text and **beyond**

**1** Comprehension check

Look at the map of the Robertsons' journey from the Galapagos Islands. What happened at places 1-8? Using information from Parts One and Two, write a sentence for each place.

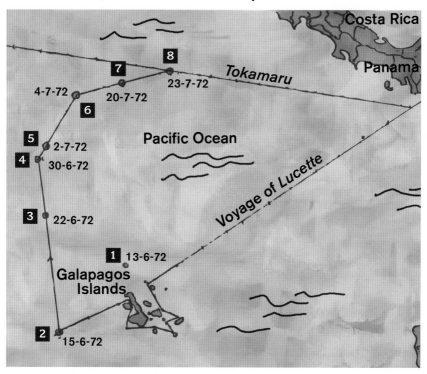

Costa Rica

**8**

**7** Tokamaru

4-7-72    23-7-72    Panama

20-7-72

**6**

**5** 2-7-72    Pacific Ocean

**4** 30-6-72

Voyage of Lucette

**3** 22-6-72

**1** 13-6-72

Galapagos
Islands

**2** 15-6-72

1 .................................................................................................

2 .................................................................................................

3 .................................................................................................

4 .................................................................................................

5 .................................................................................................

6 .................................................................................................

7 .................................................................................................

8 .................................................................................................

**PET ② Comprehension check**

Read these sentences about Part Two and decide if each sentence is correct or incorrect. If it is correct, mark A. If it is not correct, mark B.

A  B

1   During the storm of 6 and 7 July  they were
    so happy they sang songs.   ☐ ☐

2   Dougal and Lyn quarrelled about the hot weather.   ☐ ☐

3   Dougal was angry with Douglas and Robin because
    they did not catch the turtles.   ☐ ☐

4   After thirty-six days everybody looked primitive and dirty,
    but their situation was not hopeless.   ☐ ☐

5   On the evening of day thirty-eight Dougal saw a ship
    in the distance.   ☐ ☐

6   The two flares worked but the ship did not see them.   ☐ ☐

7   The Robinsons were so weak they could not climb onto
    the ship and walk around.   ☐ ☐

8   They did not sleep well because they ate too much.   ☐ ☐

9   On the ship they did not have enough food to eat,
    so they were always hungry.   ☐ ☐

10  Afterwards, Dougal missed the hard, dangerous battle
    against the sea.   ☐ ☐

**❸ Animals**

Use the words given to write four sentences in the past tense (過去式) about the animals in the story. There is an example at the beginning (0).

0   attack/sink the *Lucette*/9.55 am/June 15th
    .At 9.55 am on June 15th whales attacked and the Lucette sank..

1   third day/fall in dinghy/eat it for breakfast
    ......................................................................................................

2   all day and night/swim against raft/sometimes bite them
    ......................................................................................................

**3**  6 July/catch/a hundred eggs in it

........................................................................................................................

**4**  Dougal/hit/paddle/swim away

........................................................................................................................

**4 What did they do?**

What did the Robertsons do with the objects in the pictures? Match the sentences to the correct pictures. Write 1-6 in the boxes A-F.

**A** ☐   **B** ☐   **C** ☐

**D** ☐   **E** ☐   **F** ☐

1  Dougal cut up the turtles and fish with it.
2  Dougal lit it to signal to the ship.
3  Everybody moved to it when the raft's condition got worse.
4  Dougal hit a shark with it.
5  They always talked about it.
6  Dougal tried to catch a shark with it.

**Now write the names of the objects.**

A  ...........................   C  ...........................   E  ...........................

B  ...........................   D  ...........................   F  ...........................

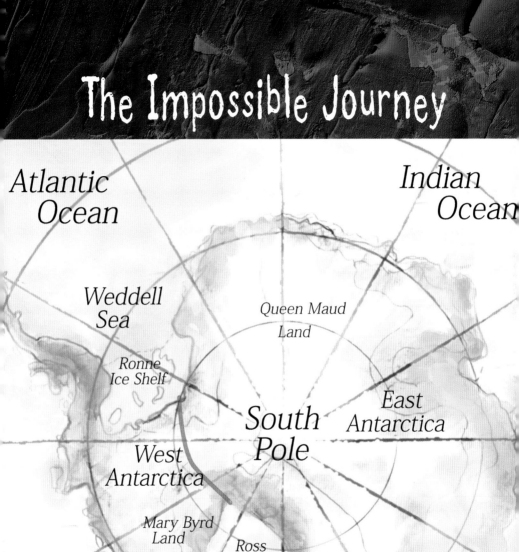

# The Impossible Journey

Atlantic
Ocean

Indian
Ocean

Weddell
Sea

Queen Maud
Land

Ronne
Ice Shelf

East
Antarctica

South
Pole

West
Antarctica

Mary Byrd
Land

Ross
Ice Shelf

Ross
Sea

Pacific
Ocean

# Before you read

**1** **Antarctica**

Look at the map on page 37. Find your friends to talk about Antarctica. Make a list of three facts that you are sure about and three things that you think are true but you are not completely sure of. Exchange your ideas. Use reference books or the Internet to check what you are not sure of.

**2** **Prediction**

You are going to read about an expedition to Antarctica in 1914-16. What do you think were the main difficulties that expeditions to Antarctica faced in those days?

**3** **Listening**

Listen to the beginning of Part One and decide if each sentence is correct or incorrect. If it is correct, put a tick (✓) in the box under A for YES. If it is not correct, put a tick (✓) in the box under B for NO.

|  | A | B |
|---|---|---|
| 1 Shackleton showed his drawing of Antarctica to a man sitting next to him. | ☐ | ☐ |
| 2 Mr Bell thought Shackleton's plan was possible. | ☐ | ☐ |
| 3 Shackleton dreamed of winning a lot of money by gambling. | ☐ | ☐ |
| 4 Shackleton hoped to cross Antarctica at its narrowest point. | ☐ | ☐ |
| 5 Mr Bell thought Shackleton's journey was going to be easy. | ☐ | ☐ |
| 6 Shackleton knew how many men he needed. | ☐ | ☐ |
| 7 Shackleton planned to sail from England to the Antarctic. | ☐ | ☐ |
| 8 The men in the second group had to leave food and water on the Beardmore Glacier for Shackleton's group. | ☐ | ☐ |
| 9 The last part of Shackleton's journey was from the South Pole to the Ross Sea. | ☐ | ☐ |
| 10 Somebody had already crossed the Antarctic on foot before Shackleton. | ☐ | ☐ |

## PART ONE

# The End of a Great Dream

One evening at a dinner party Ernest Shackleton drew a map of Antarctica on a menu. He showed it to his neighbour and said, 'Look, Mr Bell. I'm going to cross the Antarctic on foot by a route across the South Pole.'

Mr Bell was very surprised.

'That's an incredible [1] plan, Mr Shackleton,' he said.

'I know, but it's my great dream,' Shackleton answered.

An Irishman, Ernest Shackleton was a dreamer. He also had the spirit of a gambler [2], always ready to take a chance. This made him a great adventurer.

'How are you going to do it?' Mr Bell asked.

'I'm going to start from here the Weddell Sea and finish at the Ross Sea here, so I'll cross Antarctica where it's narrowest.

---

1.  **incredible**：不可思議的。      2.  **gambler**：賭徒。

Like this.' And Shackleton drew a line across the map on the menu.

Mr Bell smiled. 'It looks easy on paper. How many men will you need?'

'I don't know yet, but there will be two groups of men. The first group will sail with me from London to the Weddell Sea, and we'll start the journey. The other men will sail from Tasmania to the Ross Sea. They will put some supplies of food and water on the Beardmore Glacier [1]. At the end of our journey we'll have to come down this glacier from the South Pole. If we succeed, it will be the first journey on foot across the Antarctic.'

Mr Bell smiled again and said, 'I hope you succeed. Good luck, Mr Shackleton!'

Shackleton's ship *Endurance* [2] sailed south from London in August 1914. When it left South Georgia Island in December, Shackleton knew that the ice in the Weddell Sea was very bad. After two days the *Endurance* was sailing through heavy ice. On 19 January 1915 the ship became stuck in the ice. And by the middle of February it was frozen into a huge piece of ice, almost five square kilometres, which was moving slowly northwest, away from Shackleton's destination [3].

Three months later, he wrote: 'The ice is squeezing the ship.'

At the end of October, after ten months frozen in the ice, Shackleton said to his men, 'We must move all the boats, sledges [4] and supplies onto the ice. Tomorrow we're going to abandon the ship.'

---

1. **glacier** : 冰川。

2. **Endurance** : 此處是船名，意思是耐力。

3. **destination** : 目的地。

4. **sledges** : 雪橇。

On 27 October the ice pushed the ship up like a child's toy and crushed[1] it. The *Endurance* sank on 21 November 1915.

Now Shackleton and his men were left on the ice. The nearest land was Paulet Island, 552 kilometres away. So on 23 December they set out to reach it. But after seven days of hard work they had travelled only twelve kilometres, because the ice was moving in the opposite direction!

Shackleton decided to camp on the ice. Three and a half months later, he was still there.

'Our supplies are finished,' he wrote. 'We shoot penguins and seals for food. I have decided to try and get to Deception Island. Sometimes whaling ships[2] stop there.'

So in April 1916 they set off in the lifeboats. On the first night they camped on the ice, but it broke in two under one of their tents. Shackleton pulled a man out of the water only seconds before the ice closed again. The man was still in his sleeping bag!

On 15 April, when the wind changed to the opposite direction, Shackleton decided to change his plan too. He told his men, 'Elephant Island is only a hundred and sixty-one kilometres away. We'll have to go there.'

Three days later they landed on Elephant Island. But they found that it was only a large, empty rock with cliffs and glaciers. There was no protection from the strong wind. When they put up their tents on a rocky beach, the wind soon tore them to pieces. The men had to live and sleep under a lifeboat.

'Nobody will find us here,' Shackleton told his men, 'and there

1. **crush** : 碾碎。
2. **whaling ship** : 捕鯨船。

are no animals for food. The nearest inhabited land is South Georgia Island, which we left in December 1914. It's 1,400 kilometres across the stormiest and most dangerous sea in the world. But we must try to reach it and get help. It's the only thing we can do.'

Shackleton chose five men for this impossible voyage. Their boat was the *James Caird*, a lifeboat from the *Endurance*. It was seven metres long and nearly two metres wide. The expedition's

carpenter[1] made a big cover from canvas[2] to shelter[3] the men. But it was only a fragile[4] shelter against the storms in the Antarctic.

On 24 April 1916, nearly two years after they had left England, the six men began their desperate journey to search for help. The rest of the men, twenty-two of them, waited on Elephant Island.

1. **carpenter**：木匠。
2. **canvas**：帆布。
3. **shelter**：庇護。
4. **fragile**：脆弱的。

# The text and **beyond**

**1** **Comprehension check**

Look at the map of Shackleton's route from South Georgia to Elephant Island. Match the sentences (1-8) with the dates (A-H).

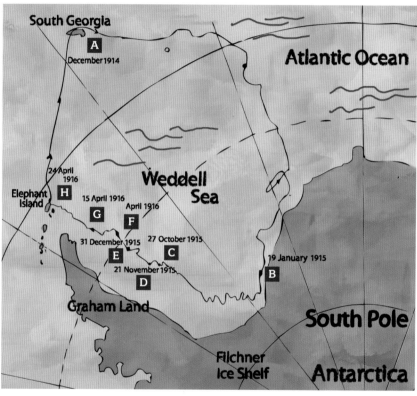

1 ☐ Shackleton decided to camp on the ice.
2 ☐ The *Endurance* was crushed.
3 ☐ Shackleton and his men set off for Elephant Island.
4 ☐ The *Endurance* sailed for Antarctica.
5 ☐ Shackleton and five men began the journey to South Georgia.
6 ☐ The *Endurance* sank.
7 ☐ The *Endurance* first became stuck in the ice.
8 ☐ Shackleton and his men tried to reach Deception Island.

## 2 Comprehension check

**Answer the questions.**

1 Why did Shackleton want to cross the Antarctic on foot?

2 When did the *Endurance* first become stuck in the ice?

3 What was Shackleton's situation in mid-February 1915?

4 In what way did Shackleton's great dream end?

5 Why did Shackleton decide not to try and reach Paulet Island?

6 What kind of help did Shackleton hope to find at Deception Island?

7 How did Shackleton save a man's life on the way to Deception Island?

8 What two things made Shackleton decide to go to Elephant Island?

9 What sort of place was Elephant Island?

10 Why was the journey from Elephant Island to South Georgia almost impossible?

T: GRADE 6

## 3 Speaking: travel and adventure

**Ernest Shackleton was a dreamer and a great adventurer. Talk about your travel dreams and intentions. Use the points below to help you.**

1 Where do you intend to go and what do you intend to do for your next holiday?

2 Which type of adventure would you prefer: a trip to a deserted island in the sun or an organised holiday to the pyramids? Why?

3 If you were able to travel anywhere, what would be your dream destination and why?

PET **4** The Antarctic climate

Look at the sentences below about the Antarctic climate. Read the text and decide if each sentence is correct or incorrect. If it is correct, mark A. If it is not correct, mark B.

**A  B**

1  Antarctica is almost as large as the Arctic.  ☐ ☐

2  The weather is very cold in the winter and very hot in the summer.  ☐ ☐

3  Life in the interior regions is very difficult.  ☐ ☐

4  Antarctica is surrounded by an ocean.  ☐ ☐

5  The water makes the climate milder especially in central regions.  ☐ ☐

6  It often snows in the coastal areas of the continent.  ☐ ☐

7  It is sunny for long periods of time in northern areas.  ☐ ☐

8  In the Antarctic Peninsula it rains less than the rest of the continent.  ☐ ☐

Antarctica is the most southern continent and it covers about 12 percent of the surface of the globe, an area twice as large as that of the Arctic.

The weather and climate of Antarctica are unique: temperatures reach a minimum of between -80°C and -90°C (-112°F and -130°F) in the interior and reach a maximum of between +5°C and +15°C (41°F and 59°F) near the coast in summer.

Antarctica is characterized by three climatic regions: the interior, the costal areas and the Antarctic Peninsula. In the interior extremely low temperatures make life impossible. These temperatures are due to a very high altitude, several months of complete darkness, fierce winds and blowing snow. Unlike the Arctic region, Antarctica is a continent surrounded by an ocean which means that the interior areas do not benefit from the moderating influence of water.

The coastal areas have milder temperatures and much more rain, most of which falls as snow, depending on location. Further north, long periods of continued sunshine occur during the summer.

The Antarctic Peninsula extends further north than the rest of the continent and it has a warmer and wetter climate than the coastal areas. Here life is more abundant than in the rest of the continent. Birds and marine mammals [1] nest and breed, and some vegetation [2], in the form of grasses and lichens [3], can be found.

1. **mammals**：哺乳類。
2. **vegetation**：植物。
3. **lichens**：地衣。

**5 Writing**

Here is part of a letter you have received from a friend.

> Before I'm too old I'm going to ride across the desert on a camel. I want to go to America and see the Grand Canyon, and these are only two of my great dreams! What about you? Have you got any great dreams?

Write a letter of about 100 words answering your friend's question.

# Before you read

**1** Prediction

A Will Shackleton and his companions reach South Georgia Island? Choose one of the following.

1 ☐ No, it's impossible.
2 ☐ Yes, but only after many dangers.
3 ☐ Yes, but some of them will die on the way.

B What kind of dangers will they meet? Use the verbs in the box to write a sentence about the possible consequences of the dangers 1-6, as in the example. You can use a dictionary.

> freeze (to death)   turn over   sink   drown
> die from exhaustion   fall overboard   lose hope
> die of thirst   become weak and inefficient

0 storms: Perhaps the boat will sink.
1 sea water freezing on the boat: ...............................................
2 enormous waves: ...............................................
3 the men always wet and cold: ...............................................
4 no rest for the men: ...............................................
5 not enough water: ...............................................

**2** Listening

Listen to the first part of Part Two and say whether the following statements are true (T) or false (F).

|  | | T | F |
|---|---|---|---|
| 1 | The ice from the sea water made the boat dangerously heavy. | ☐ | ☐ |
| 2 | The sea was always stormy, so the men could not have a good rest. | ☐ | ☐ |
| 3 | At first, Shackleton thought the gigantic wave was a white cloud on the horizon. | ☐ | ☐ |
| 4 | When the wave hit the boat, the men lost hope. | ☐ | ☐ |
| 5 | The men were very thirsty when they landed on South Georgia. | ☐ | ☐ |

## PART TWO

# Too Incredible to Be True?

Storm after storm hit the *James Caird*. Sea water froze on the boat, and the weight of the ice nearly made it capsize [1]. With waves over fifteen metres high and ninety metres long underneath them the men crawled [2] across the deck to break off the ice. Many times they thought that the boat was going to sink. They were always wet and cold and they never had a good rest because the stormy sea threw the boat around.

On the 11th day the biggest wave they had ever seen came towards them. At first, Shackleton thought it was a break in the clouds on the horizon. Then: 'I realized that I had seen the white top of an enormous wave,' he wrote later. 'I have never seen a wave so gigantic [3] in my twenty-six years at sea.'

'Hold on! Hold on! It's got us!' he shouted.

The wave lifted the boat up and threw it forward like a cork [4]. The men fell around inside as the wave crashed over it. The boat was full of freezing water, but the men continued to throw the

---

1. **capsize**：〔船隻〕傾覆。
2. **crawled**：爬。
3. **gigantic**：巨大的。
4. **cork**：軟木塞。

water out with anything they could find. They knew they were fighting for their lives. After ten minutes the *James Caird* seemed to recover from the attack. It had survived!

During the following days the supply of water got very low. The men were very thirsty and their tongues felt big. On 11 May, fourteen days after they had left Elephant Island, they saw the coast of South Georgia Island. Landing on a beach, they found a stream of fresh water. They drank and drank the icy water. It was a wonderful moment!

But their journey was not over. They had landed on the south of the island, 241 kilometres away by sea from Stromness, the whaling station on the north side.

'Men, the *James Caird* probably won't survive another voyage,' Shackleton said. 'And we are all too exhausted [1]. So I've decided to take two of you with me and walk across the island to the station.'

They started out on 19 May, taking a compass, a stove [2], a tent, fifteen metres of rope and supplies for three days. Nobody had ever walked across the mountains of South Georgia before and Shackleton and his companions did not know exactly where the whaling station was. This was the second part of an impossible journey!

On the first day they climbed a glacier in thick fog. When the fog lifted a little, there was a deep crevasse [3] just in front of them! They had seen it just in time. Time after time they climbed the frozen mountains and found deep valleys or cliffs in their way. Finally, they reached a cliff and saw below them only thick fog. They knew they had to find a place soon to put up their tent

1. **exhausted** : 筋疲力盡。
2. **stove** : 爐。
3. **crevasse** : 冰隙。

for the night. There was only one thing to do: slide down the mountain! Using the ropes, they made three seats. Each man sat on one and put his legs round the man in front. Shackleton was the first man. He pushed himself down into the fog.

Afterwards, he wrote, 'We seemed to fly into space. My hair stood on end. Then suddenly I knew I was enjoying it. We were shooting down the side of a vertical mountain at one and a half kilometres every minute. I shouted with excitement and the other men shouted too. Our toboggan [1] stopped safely in the soft snow.'

Now there was only one more obstacle [2]. As they were walking through a narrow valley at the bottom of a mountain, they came to a waterfall. The water fell about nine metres, with cliffs on each side. They had to climb down in the waterfall. Using the ropes again, they tied one end over a big rock, and each man went down through the water. They were very wet and cold when they reached the bottom, but safe!

At last, on 21 May, Shackleton and his companions walked into the whaling station. Everyone was amazed. Their 1,600-kilometre journey seemed impossible. It was too incredible to be true!

Immediately, some of the men at the station sailed to the south of the island to rescue the other three men. But the twenty-two men on Elephant Island had to wait four months before they were rescued. At the end of September a ship managed to break through the ice around the island. The men were very weak and almost dead. They had been twenty weeks on the island during the Antarctic winter. It was a miracle that they all survived.

Although the expedition had failed, the courage and endurance of the men were extraordinary [3]. It had been a heroic adventure.

1. **toboggan**：平底雪橇。
2. **obstacle**：障礙。
3. **extraordinary**：非凡的。

## The text and **beyond**

**PET** **1** **Comprehension check**

**For questions 1-6 choose the correct answer — A, B, C or D.**

1   The great wave threw the boat forward like a cork and
   A ☐ sank it.
   B ☐ filled it with freezing water.
   C ☐ the men fell into the sea.
   D ☐ smashed it to pieces.

2   Shackleton and his men landed on the island
   A ☐ a long way from the whaling station.
   B ☐ and their journey was over.
   C ☐ on a beach on the north side.
   D ☐ and had to walk 241 kilometres to Stromness.

3   'There was only one thing to do.' What was it?
   A ☐ Put up their tent.
   B ☐ Go back the way they had come.
   C ☐ Climb down the mountain.
   D ☐ Slide down the mountain sitting on ropes.

4   At the bottom of the waterfall the men were
   A ☐ safe and dry.
   B ☐ wet but safe.
   C ☐ cold and dry.
   D ☐ wet and in danger.

5   When were the men on Elephant Island rescued?
   A ☐ On 21 May 1916.
   B ☐ Twenty weeks after Shackleton reached Stromness.
   C ☐ In September 1916.
   D ☐ Four months after the *Endurance* sank.

6   At Stromness everybody was amazed because

A ☐ they did not believe Shackleton's story.

B ☐ Shackleton and his companions had travelled 1,600 kms across the mountains.

C ☐ Shackleton's expedition had succeeded.

D ☐ Shackleton's journey was incredible but true.

**2** Shackleton and the commander

**When Shackleton walked into Stromness, what did the commander of the whaling station say? Complete the dialogue.**

**Commander**: Who (**1**) ..................................... ?

**Shackleton**: Hello, Commander. I'm Ernest Shackleton.

**Commander**: Shackleton? I don't believe you!

**Shackleton**: (**2**) ..................................... ?

**Commander**: Because I don't recognize you. Are you the same man who left here eighteen months ago in the *Endurance*?

**Shackleton**: Yes. And here are Mr Worsley and Mr Crean. Don't (**3**) ..................................... them either?

**Commander**: Yes, now I do. But what (**4**) ..................................... here? Where's your ship?

**Shackleton**: The *Endurance* was (**5**) ..................................... on 27 October and sank last November.

**Commander**: What! So how (**6**) ..................................... here?

**Shackleton**: We came by sea from Elephant Island, and then we (**7**) ..................................... .

**Commander**: But that's too (**8**) ..................................... true! Where are the rest of the men? Are they dead?

**Shackleton**: I hope not. (**9**) ..................................... on Elephant Island. And three men are waiting on a beach on the south of this island.

**Commander**: (**10**) ..................................... a rescue boat immediately! Welcome, Shackleton! If your story is true, it's really amazing!

**3** **Antarctica**

Read the text below and choose the correct word (A, B, C or D) for each space. There is an example at the beginning (0).

Antarctica is the coldest place (0) .....B.............. the world. In winter the temperature (1) ..................... sometimes fall to -70°C or (2) ..................... . Covering 12,100,000 square kms, Antarctica is the fifth largest (3) ..................... on earth. 98% of the Antarctic is under (4) ..................... and ice, and in (5) ..................... places the ice is 3,500 m thick. The weight of the ice (6) ..................... it move up to 200 m (7) ..................... year. West and East Antarctica are separated (8) ..................... the Transantarctic Mountains. These are as high (9) ..................... 3,000 m, (10) ..................... Antarctica is also the world's highest continent.

| | | | | | | | |
|---|---|---|---|---|---|---|---|
| 0 | **A** on | **B** in | | **C** at | | **D** of | |
| 1 | **A** is | **B** must | | **C** can | | **D** would | |
| 2 | **A** lower | **B** higher | | **C** larger | | **D** deeper | |
| 3 | **A** state | **B** nation | | **C** continent | | **D** country | |
| 4 | **A** forest | **B** water | | **C** cloud | | **D** snow | |
| 5 | **A** some | **B** every | | **C** any | | **D** which | |
| 6 | **A** cause | **B** makes | | **C** lets | | **D** has | |
| 7 | **A** one | **B** a | | **C** the | | **D** last | |
| 8 | **A** with | **B** of | | **C** on | | **D** by | |
| 9 | **A** like | **B** to | | **C** as | | **D** over | |
| 10 | **A** but | **B** so | | **C** which | | **D** then | |

**4** **Shackleton's journal**

Complete the sentences from Shackleton's journal with the words in the box in their correct form.

| survive | thirsty | tongue | gigantic | stream | crawl | cold |
|---|---|---|---|---|---|---|

**7 May** To break the ice we have to (1) ..................... across the deck in case we fall into the sea.

**8 May** The most (2) ..................... wave I've ever seen hit us, but the boat has (3) ..................... .

**9 May**    Not much water left, so we are **(4)** ..................... .

**10 May**    Our **(5)** ..................... feel very big in our mouths.

**11 May**    South Georgia at last! We landed on a beach where there was
a **(6)** ..................... . The water was as **(7)** ..................... as ice,
but it was wonderful!

**5** Another great explorer: Marco Polo

**Read the text below about Marco Polo. Fill in the gaps with the words
in the box.**

return    journey    service    home    Wall

jewels    ship    Venice    deserts

Marco Polo was fifteen when his father Nicolò and his uncle Maffeo
returned to Venice after a **(1)** ..................... of thousands of kilometres
across mountains and **(2)** ..................... to the court of Kublai Khan near
Peking. When young Marco heard their fabulous stories, he wanted to
go with them on their **(3)** ..................... journey. In 1271 the brothers left
for China with Marco.

First they travelled down to the Persian Gulf, but then decided not to
continue by **(4)** ..................... . They went north towards Asia, crossed
the great Gobi Desert and finally came to the Great **(5)** .....................
of China. Their journey took three and a half years.

The Grand Khan received them warmly as old friends. He was
especially happy to see Marco. They worked in the Khan's
**(6)** ..................... and became rich. But after seventeen years they
wanted to go back to **(7)** ..................... . Kublai Khan was not happy
about this, but finally he let them go.

So in 1292 they sailed to Cambodia, Borneo, Sumatra, Sri Lanka, India
and in 1294 they arrived in the Persian Gulf. Then the Polos set out for
**(8)** ..................... . At the end of 1295 they arrived in Venice. At first,
their relatives did not recognize them and refused to believe they were
from the Polo family until they saw all the precious **(9)** .....................
and presents from the Grand Khan!

**6** Discussion

**Can you think of any other famous explorers? What do you remember
of their journeys?**

T: GRADE 6

**7** **Speaking: travel and tourism**

Since Shackleton's time things have changed in Antarctica and the Arctic. Today tourists regularly visit them on cruise ships. In the Arctic there are many nuclear power stations, missile bases and industrial sites for oil and minerals. More and more people travel to deserts, lonely islands like the Galapagos, jungles and mountain regions like Tibet and Nepal. Find some information or pictures that show travel and tourism to the world's isolated regions. Talk about the places, using the questions below to help you:

- where is the place and how long have tourists been visiting it?
- what do builders of hotels, shops and roads in isolated areas have to think about?
- what do tourists to lonely islands and areas of nature and wildlife need to think about?
- if too many tourists visit wild and beautiful places what will happen to the local people and traditions?

**INTERNET** PROJECT

Connect to the Internet and go to www.blackcat-cideb.com or www.cideb.it. Insert the title or part of the title of the book into our search engine. Open the page for *True Adventure Stories*. Click on the Internet project link. Go down the page until you find the title of this book and click on the relevant link for this project.

1  Find out more about Antarctica.
- ▶ What is the one important difference between Antarctica and the Arctic?
- ▶ How long has the Antarctic been covered with ice?
- ▶ What was it like over 100 million years ago?

2  Organize your class into two groups. Each group can report to the class on two of the following subjects:
- ▶ Captain Cook's voyage in the 18th century
- ▶ 19th century exploration
- ▶ the race to the South Pole (Amundsen and Scott)
- ▶ the wildlife of Antarctica.

# The Big Heat

In 2002, 3,240 square kilometres of ice broke from Antarctica and disintegrated [1]. Frozen in the Antarctic ice is 70% of the world's fresh water. But since the 1970s the ice has been melting, and if all of it melts, Earth's oceans will rise almost 70 metres.

In November 1988 the sea in James Bay in the Arctic did not freeze. That winter it froze six weeks later than usual for the first time.

The melting ice in Antarctica and the unusually warm winter in the Arctic are the consequences [2] of global warming. This is a rise in the

1. **disintegrated**：崩解。　　　2. **consequences**：結果。

Earth's temperature when the levels of carbon dioxide ($CO_2$) and other natural gases in the atmosphere increase. These gases are like a warm 'greenhouse' over our planet. Without them the Earth would be too cold for life. The increase in greenhouse gases like $CO_2$ is called the 'greenhouse effect [1]'. It has happened many times during climate changes in Earth's history. But today it is happening much faster than before because human activity has added more $CO_2$ to the atmosphere. In the past, changes in climate took hundreds of thousands of years, so nature had time to adapt to them. But today's global warming [2] is happening ten times faster. During the last 150 years industrial activity has increased $CO_2$ by 30%, and so the Earth's temperature has risen.

The modern industrial world burns fossil fuels [3] like petrol, coal and natural gas, which create more $CO_2$. By the year 2050 the level of $CO_2$ will be twice that in the atmosphere before industrialization began. And many scientists say the temperature will rise another 5° C by the year 2100.

1. **greenhouse effect**：溫室效應。　　2. **global warming**：全球變暖。
3. **fossil fuels**：化石燃料。

As we continue to burn carbon fuels in our cars, factories and electric power stations and as we destroy more forests for agriculture, global warming is already changing the weather on our planet. In the next decades more – or less – rain will cause floods and droughts. If they damage farming, food will become harder to produce. Then there will be illness, disease and famine [1]. Sea levels will rise from the melting polar ice and cause disaster [2] to communities on the coasts, where perhaps some cities will disappear. There will be more storms and hurricanes. The oceans will become warmer, so animals and plants will have to adapt or become extinct.

But a small number of scientists do not think human activity has caused a dangerous global situation. In their opinion, it is not the increase in greenhouse gases that makes a big difference to the climate. They believe that the sun's activity is the biggest influence

1. **famine** : 饑荒。   2. **disaster** : 災難。

on climate. And they say that increases in the levels of $CO_2$ in the climate changes of the past always followed the global warming: the warmer temperatures caused the increase in $CO_2$, not the other way. But there is one thing all the scientists and experts agree about: the Big Heat has begun!

**1** **Comprehension check**

**For questions 1-5 choose the answer (A, B or C) which you think fits best.**

1  If all the ice in Antarctica melts
   A  ☐  the level of the world's fresh water will rise by 70 metres.
   B  ☐  sea levels will rise.
   C  ☐  winters will be warmer.

2  $CO_2$ is called a 'greenhouse gas' because it
   A  ☐  makes the Earth too cold for life.
   B  ☐  is used in greenhouses.
   C  ☐  helps to keep the earth warm.

3  Why is the global warming of today different?
   A  ☐  It has not happened before.
   B  ☐  Nature has time to adapt to it.
   C  ☐  It is happening faster than before.

4  In the next decades global warming will
   A  ☐  bring floods and droughts to many areas.
   B  ☐  increase food production.
   C  ☐  destroy all life in the sea.

5  A small number of scientists think that global warming is
   A  ☐  the cause of the greenhouse effect.
   B  ☐  caused by human activity.
   C  ☐  caused by increases in the level of $CO_2$.

**INTERNET** PROJECT

Connect to the Internet and go to www.blackcat-cideb.com or www.cideb.it . Insert the title or part of the title of the book into our search engine. Open the page for *True Adventure Stories*. Click on the Internet project link. Go down the page until you find the title of this book and click on the relevant link for this project.

1 Discuss some ways by which we can all help to combat global warming in our daily lives. Think about the following:
   ▶ electronic gadgets in the home (computers, TVs, DVD videos, standby lights, etc.)
   ▶ domestic heating (including hot water for showers, baths, dishwashers, washing machines, etc.)
   ▶ travelling (cars, aeroplanes).

2 Do you think people can change their old habits and style of living to help consume less energy? Can you?

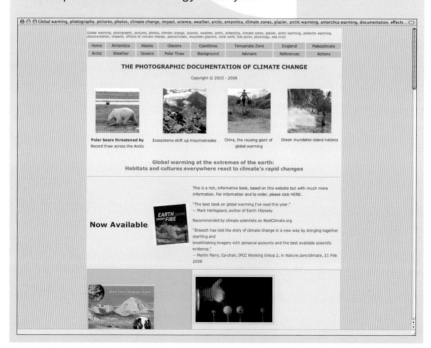

# Amy Johnson, Woman Pioneer

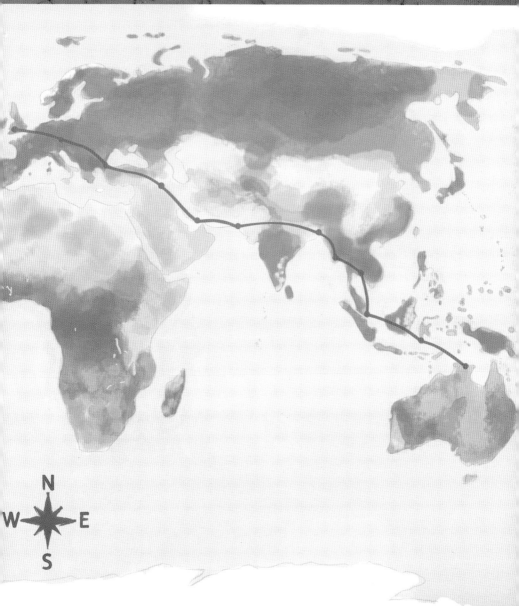

## PART ONE

# A Dream Come True

**Lord Wakefield looked at the young woman in surprise.**

'Did I hear you correctly, Miss Johnson? You say you want to fly solo [1] to Australia?'

'Yes.'

Lord Wakefield looked again at the twenty-five year old woman in front of him. She was small and had fair hair and blue eyes.

'And what qualifications [2] have you got?'

'Well, I learned to fly at the London Aeroplane Club. I took the test in July and got a pilot's A licence. I'm also a qualified mechanic [3].' The woman smiled at Lord Wakefield's expression of surprise. 'I'm the only woman in the world with an engineer's licence. And I bought my own aeroplane: a De Havilland Moth. It's not new, of course.'

'Hm. And how much flying have you done, Miss Johnson?'

Amy Johnson looked directly in his eyes. 'I've never flown more than two hundred miles, but...'

1. **solo** : 單獨地。
2. **qualifications** : 資格。
3. **mechanic** : 機械工。

66

'Have you ever flown across the English Channel?'

'N-no, I haven't, but...'

'I'm sorry, Miss Johnson,' Lord Wakefield interrupted [1] again, 'but I can't help you. The idea is impossible. An amateur [2] pilot can't make such a lonely and dangerous journey.'

Amy was disappointed, but she was determined [3] to get the money for her dream to fly from England to Australia. Nobody took her seriously, including the newspaper editors. The year was 1929. They knew the public were losing interest in heroic solo flights. The American aviator Charles Lindbergh had already flown across the Atlantic. The editors told Amy her idea was not very original. It was just not sensational [4] enough.

Amy went to the Australian Minister for Trade, who was visiting London.

He said, 'Young lady, if you want to go to Australia, I advise you to go by ship.'

Next she tried the High Commissioner for Australia, with the same result.

Then she went to see the Director of Civil Aviation, who had already helped adventurous amateur pilots. He asked Lord Wakefield to pay for Amy's fuel supplies, about £300, which was a lot of money in 1929. This was good news.

Now Amy went to the office of *The Daily News*.

'I'm going to fly solo to Australia,' she told the editor, 'because it's a great challenge and I love flying. I don't want publicity, but I've heard that when I return my experience will be useful to other pilots. I'm going to begin the journey in May 1930.'

But next day the report in *The Daily News* said that Amy

1.  **interrupted** : 打擾。
2.  **amateur** : 業餘的。
3.  **be determined** : 決心。
4.  **sensational** : 引起轟動的。

wanted to beat Captain Bert Hinkler's record to Australia. The editor knew this could sell more papers. Amy was very angry.

'The newspaper report isn't true,' she said. 'I don't want to fly fast and beat Hinkler's record.'

After that, Amy decided to stop trying to get help. She prepared for the flight.

Amy's aeroplane was light, safe and easy to fly. She had it painted green, her favourite colour, and called it *Jason*. She studied navigation. She also learned how to use a parachute [1] in half an hour! But she did not do a practice jump.

At eight o' clock on the morning of 5 May 1930 she took off secretly from Croydon Airport near London. Only her father and a few friends waved her goodbye.

First, she flew to Vienna, 1,288 kilometres away, in ten hours. Next day, after flying the same distance in twelve hours, she arrived in Istanbul. Now the world began to notice the courageous [2] girl-flyer. On the third day she landed at Aleppo, 885 kilometres away in modern Syria. Everything was going well. But on day four there were problems.

As Amy was flying across the desert to Baghdad, there was a great sandstorm [3]. She had to land and wait for the storm to pass with a gun in her hand! In this wild region the nomads [4] sometimes attacked strangers. But two hours later the storm passed and Amy arrived in Baghdad. Now the world was following the journey with enormous interest because this young female aviator had flown faster than Bert Hinkler.

1. **parachute**：降落傘。
2. **courageous**：無畏的。
3. **sandstorm**：沙塵暴。
4. **nomads**：遊牧民。

# The text and **beyond**

**1 Comprehension check**

**Answer these questions.**

1 Lord Wakefield was surprised because Amy wanted to fly solo to Australia. What surprised him a second time?
2 Lord Wakefield told Amy he could not help her. Why?
3 Why did the newspaper editors think Amy's project had to be more original and sensational?
4 In your opinion, why did Amy take off from Croydon secretly?
5 Where did Amy have to land on the way to Baghdad?
6 Why did Amy wait with a gun in her hand during the storm?

**2 Places**

**Match the sentences (1-4) to the photos (A-D).**

A
B
C
D

1 Amy wanted to fly here from England.
2 She landed here on 5 May 1930 at about six in the evening.
3 It took her twelve hours to fly here from Vienna.
4 Amy took off secretly from here.

### ❸ Characters

**Which of the people (A-H) do questions 1-10 refer to? Write A, B, C, etc. in the boxes. You will need to use two of them more than once.**

| | | | |
|---|---|---|---|
| **A** | Amy Johnson | **E** | Lord Wakefield |
| **B** | Australian Minister for Trade | **F** | Captain Bert Hinkler |
| **C** | Newspaper editors | **G** | Desert nomads |
| **D** | Director of Civil Aviation | **H** | Father and friends |

Who...

1 `F` had already made a record flight to Australia?
2 ☐ thought Amy's project would not interest the public?
3 ☐ often helped amateur pilots?
4 ☐ were at the airport when Amy began her journey?
5 ☐ was small, with fair hair and blue eyes?
6 ☐ sometimes attacked strangers?
7 ☐ told Amy her dream was impossible?
8 ☐ told Amy that she should go to Australia by ship?
9 ☐ liked the colour green?
10 ☐ took longer than Amy to fly to Baghdad?

### ❹ Word formation

**Complete the sentences with the correct form　(正確形態）of the words in brackets. There is an example at the beginning (0).**

0 The ..Australian.. Minister for Trade visited London in 1929. (*Australia*)

1 Lord Wakefield asked Amy what .................... she had. (*qualify*)

2 Amy told Lord Wakefield she was a .................... mechanic. (*qualify*)

3 Amy said, 'I'm the only woman in the world with an .................... 's licence.' (*engine*)

4 Amy told the editor of the *Daily News* that she did not want .................... . (*public*)

5 To fly alone from England to Australia in 1930 was very .................... , so Amy was a .................... young woman. (*danger/courage*)

6 In the end Amy got what she wanted because she had a lot of .................... . (*determine*)

# Bert Hinkler:
## *the Lone[1] Eagle*

On 8 December 1892 Herbert Louis John Hinkler, known as Bert, was born in Bundaberg, a town in Queensland, Australia. Flying fascinated [2] him from an early age. As a boy he studied the flight of birds. Later, he tried to fly by fixing wings to his back like Icarus in the Greek myth. In 1911 and 1912 he built his own gliders [3] and flew them up to 10 metres high on a beach near Bundaberg. Determined to make a career in aviation, he went to England in 1913. He could not pay for his passage [4], so he worked on the ship. All his life Bert Hinkler never had enough money because he was not good at 'selling' himself. He was a quiet, modest man who always liked to fly and work alone. When he became

**Bert Hinkler** photographed in about 1920.

1. **lone** : 孤寂的。

2. **fascinated** : 吸引。

3. **gliders** :  滑翔機。

4. **passage** : 航程。

72

famous, he was called 'the Lone Eagle'. One newspaper said he was 'the pilot the world forgot.'

After many disappointments Hinkler got a job with Sopwith, an aircraft company. Then in World War One he served in the Royal Naval Air Service. He received the Distinguished Service Medal in 1917 for shooting down German airplanes. After the war he worked as a test pilot in Southampton. Then in 1920 he set out to fly solo from London to Australia. He reached Turin, Italy, in less than ten hours, a new world record for a light airplane. But he had to abandon the adventure because there was war in the Middle East.

In the 1920s Hinkler created many records: for example, flying 1,900 kilometres non-stop from England to Riga in Latvia. And in 1921 he made a record-breaking flight of 1,370 kilometres from Sydney to Bundaberg.

It was February 1928 when Bert Hinkler took off on the great adventure that made his name famous, the first solo flight from England to Australia. Before Hinkler, the record was a flight of 28 days. His first stop was Rome, which he reached non-stop in a day, a new record. Then on to North Africa and Basra in Iraq. The newspapers only noticed Hinkler's journey when he reached India in the record time of seven days. Navigating perfectly across Asia and the Timor Sea using an atlas [1], he arrived in Darwin on 22 February. His fifteen-and-a-half day solo flight was the longest and fastest ever made.

But Hinkler could not stop there. He made another pioneering solo flight in 1931. He flew from New York 2,400 kilometres non-stop to Jamaica. Then he continued to Venezuela and Brazil and across the

1.  **atlas**：地圖集。

**Bert Hinkler** lands in England on 7 December 1931.

South Atlantic for West Africa. During the whole crossing of 16,000 kilometres the weather was very bad, with storms, heavy rain, cloud and fog. Using only a compass, Hinkler navigated with incredible accuracy [1] and landed only 160 kilometres from the place where he had planned to stop. Then he continued his flight to England – and another record!

In 1933 Hinkler decided to fly solo to Australia for a second time. He wanted to break the new flying record of eight days ten hours. On 7 January he took off in secret; as always, he did not want any publicity. His airplane was seen over the Italian Alps near the

---

1. **accuracy**: 準確。

Pratomagno Mountains. Then it disappeared. Months later, on 1 May, Hinkler's body was found near the wreck [1] of his airplane in the Appennines. It had crashed into the mountains, probably on 8 January. The Lone Eagle had flown for the last time.

1.  **wreck** : 飛機殘骸。

### 1 Comprehension check

**Complete the sentences.**

1   Born in Australia in 1892, Bert Hinkler ................................. by flying from an early age.

2   For example, he used to watch birds to see ................................. .

3   Later, he attached wings ................................. and ................................. .

4   In 1913 he ................................. because ................................. .

5   When he became famous he was called ................................. because he was ................................. and he ................................. ; he did not want ................................. .

6   While serving in the First World War, he ................................. and for this he received a medal.

7   Unfortunately, his first attempt to fly solo from England to Australia in 1920 ................................. .

8   But eight years later he tried again and he ................................. . He flew ................................. .

9   He made other records in ................................. when he ................................. .

10  In 1931 he made a pioneering solo flight ................................. ; he ................................. although ................................. .

11  Then in 1933, during a second solo flight to Australia, ................................. .

# Before you read

**1 Listening**

Listen to the beginning of Part Two and decide if each sentence is correct or incorrect. If it is correct, put a tick (✓) in the box under A for YES. If it is not correct, put a tick (✓) in the box under B for NO.

|   |   | A | B |
|---|---|---|---|
| 1 | On the fifth day Amy landed at Basra in southern Iraq. | ☐ | ☐ |
| 2 | She could not land on the coast of the Gulf because there were mountains. | ☐ | ☐ |
| 3 | She got to Karachi two days after leaving Banda Abbas in Iran. | ☐ | ☐ |
| 4 | To beat Hinkler's record Amy had to fly to Australia without landing. | ☐ | ☐ |
| 5 | It took Amy eight days to fly from London to Calcutta. | ☐ | ☐ |
| 6 | When she arrived in Calcutta, she was eight days ahead of Hinkler's record. | ☐ | ☐ |
| 7 | On the way from Calcutta to Rangoon there were some very big mountains. | ☐ | ☐ |
| 8 | Amy had to fly much higher than usual to escape the storms. | ☐ | ☐ |

**2 Follow Amy's route**

Listen again, look at the map and fill in the missing information in the spaces A-D.

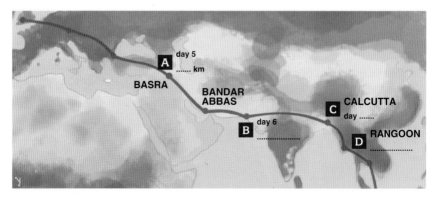

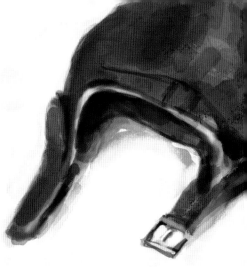

## PART TWO

# Lost and Found

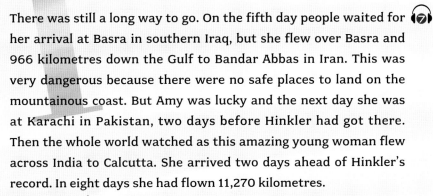

There was still a long way to go. On the fifth day people waited for
her arrival at Basra in southern Iraq, but she flew over Basra and
966 kilometres down the Gulf to Bandar Abbas in Iran. This was
very dangerous because there were no safe places to land on the
mountainous coast. But Amy was lucky and the next day she was
at Karachi in Pakistan, two days before Hinkler had got there.
Then the whole world watched as this amazing young woman flew
across India to Calcutta. She arrived two days ahead of Hinkler's
record. In eight days she had flown 11,270 kilometres.

But Amy was feeling very tired now. She knew she could not
beat Hinkler's record because he had flown the rest of the route
to Australia without stopping for a rest.

Now came one of the most difficult and dangerous moments
of her journey. Her next destination was Rangoon (now Yangon),
the capital of Burma. To get there she had to fly across the

immense [1] Yomas Mountains. When she was flying above them, storms of wind and rain battered [2] her little aeroplane and she had to descend [3] to only forty-six metres. Desperately she looked for somewhere to land.

Amy was very nervous. She could not see much because there was a lot of mist. She flew on, looking for a place to land. Finally, she saw a small football field outside Rangoon and she landed between two goal posts [4]! But she had an accident on the football field and had to wait while mechanics repaired the wing of her aeroplane. Then she could not take off from the field because it was too small, so the plane was taken to Rangoon. Two days later she left for Bangkok, over 500 kilometres away.

Although she was now behind Hinkler's record, the excited world watched and waited. On 16 May she met heavy rain and wind again. She was flying in thick cloud over jungle and mountains. She decided to climb up to 1,550 metres. Several hours later she landed near Bangkok, safe but exhausted.

On her thirteenth day Amy flew down Malasyia in bad weather, hoping to reach Singapore. She received a fantastic welcome when she arrived. Now her name was famous all over the world. But in front of her was the final and most dangerous stage of her adventure: 3,300 kilometres across Indonesia and the Timor Sea to Darwin in Australia.

She took off at six in the morning. At midday the people on the Indonesian island of Sumbawa saw her aeroplane. Then nobody saw or heard of her for the rest of the day. Night came. Still no news. Everyone waited anxiously. No message arrived. People began to think that Amy had crashed somewhere over the

1. **immense** : 極大的。
2. **battered** : 擊打。
3. **descend** : 下降。
4. **goal posts** : 球門柱。

sea. Early the next morning preparations for a search began. A ship was ordered to search the Timor Sea and flying-boats [1] were ready to go.

Suddenly the news came that Amy had landed at Haliloelik, a lonely village on an island about 19 kilometres from the Timor island of Atamboea. She had sent a telegram.

'No telephone at Haliloelik. I got transport into Atamboea and found somewhere to sleep. I'm safe.'

After this many people thought, 'Amy won't continue her journey. It's too dangerous.' But she was determined to finish it. Luckily, when she flew across the Timor Sea, the weather was good. It was the nineteenth day of her epic flight. She flew over the empty sea and the coast of northern Australia without a radio. If she lost her way again, there was not much chance of survival. But her navigation was perfect. She arrived at Darwin, where an enormous crowd was waiting.

Amy Johnson had flown alone more than 19,300 kilometres across the world. It was an amazing achievement. She received a special honour called a CBE (Citizen of the British Empire), and the Daily Mail gave her £10,000.

In the 1930s Amy continued to fly and she made more record flights. When war broke out in 1939, she worked as a pilot helping other pilots who got into difficult situations. On 5 January, 1941, she disappeared over the Thames Estuary during bad weather, and was never seen again. She was thirty-seven years old.

1. **flying-boats :**  飛艇；水上飛機。

# The text and **beyond**

**1** Comprehension check

**Match the first half of each sentence (A-L) to the second half (1-12) to make a summary of Part Two.**

A ☐ On the 5th day Amy flew

B ☐ She was two days ahead of Hinkler's record

C ☐ Then the whole world watched

D ☐ But now, after flying for eight days and 11,270 kilometres,

E ☐ And she could not beat Hinkler,

F ☐ During her flight to Rangoon she had to cross

G ☐ She finally landed on a football field near Rangoon,

H ☐ From there, she continued to Singapore,

I ☐ On the 14th day she took off at 6am

J ☐ But after passing the island of Sumbawa in Indonesia

K ☐ Then she sent a telegram from an island called Atamboea

L ☐ Finally, on the 19th day she flew to Darwin without a radio

1   as she flew on to Calcutta.

2   the dangerous Yoma Mountains in storms.

3   where the people gave her a fantastic welcome.

4   when she arrived in Karachi, Pakistan.

5   Amy was very tired.

6   she disappeared.

7   from Baghdad to Banda Abbas in Iran.

8   and left for Bangkok two days later.

9   to say that she was safe.

10   who had flown to Australia without resting.

11   to fly the 3,300 kilometres to Darwin in Australia.

12   and reached it after flying 19,300 kilometres across the world.

'In eight days she had flown 11,270 kilometres.'

We use the **past perfect** to talk about an action that happened at a time **before** another action in the past:

*People began to think that Amy **had crashed** somewhere over the sea.*
*Suddenly the news came that Amy **had landed** at Haliloelik.*

We use the past perfect with **after, although, because, before, until,** when we talk about the earlier action:

*But Amy was lucky and the next day she was at Karachi in Pakistan, two days **before** Hinkler **had got** there.*

*She knew she could not beat Hinkler's record **because** he **had flown** the rest of the route to Australia without stopping for a rest.*

We also use the past perfect with **already** and **never**:

*Then she went to see the Director of Civil Aviation, who **had already helped** adventurous amateur pilots.*

**2** **The past perfect**
**In each sentence below, fill in the gaps with one verb in the past simple（簡單過去式）and one verb in the past perfect（過去完成式）.**

1 Amy ................. to fly solo after she ................. a pilot's A licence. *(decide/get)*

2 After she ................. an aeroplane Amy ................. to talk to Lord Wakefield. *(buy/go)*

3 Lord Wakefield ................. Amy if she ................. across the English channel before. *(ask/fly)*

4 The editors of a newspaper ................. Amy that her idea wasn't original because the American aviator Charles Lindbergh ................. across the Atlantic. *(tell/fly already)*

5 Amy ................. to the Director of Civil Aviation, because he ................. adventurous amateur pilots. *(talk/help already)*

6 Amy ................. in Pakistan two days before Hinkler ................. there. *(arrive/get)*

7 Amy ................. Hinkler's record because he ................. for a rest. *(can not beat/stop)*

8 People ................. that Amy ................. somewhere over the sea. *(think/crash)*

## PET ❸ Amy Johnson's last flight

**Read the text below and choose the correct word (A, B, C or D) for each space. There is an example at the beginning (0).**

Amy Johnson disappeared over the Thames Estuary in 1941 (**0**) ..D... she was doing work for the war. She abandoned her aeroplane in the air and came (**1**) ....... by parachute. (**2**) ....... there was a mystery. She was 161 kms from (**3**) ....... destination, and two hours late. Also, witnesses said they saw two people (**4**) ....... into the water. So (**5**) ....... people believe that Amy was returning from a mission in France with a secret agent. But this idea is not very (**6**) ....... . Amy was flying in bad weather conditions without a (**7**) ....... . She probably lost her way and did not have (**8**) ....... fuel. The 'secret agent' was probably the exit door, (**9**) ....... Amy had to throw away to make the (**10**) ....... by parachute.

| 0 | **A** because | **B** so | **C** but | (**D**) while |
|---|---|---|---|---|
| 1 | **A** up | **B** down | **C** over | **D** along |
| 2 | **A** But | **B** Then | **C** Because | **D** Since |
| 3 | **A** the | **B** this | **C** a | **D** her |
| 4 | **A** swimming | **B** falling | **C** flying | **D** walking |
| 5 | **A** any | **B** all | **C** some | **D** every |
| 6 | **A** true | **B** right | **C** impossible | **D** probable |
| 7 | **A** TV | **B** telephone | **C** radio | **D** map |
| 8 | **A** enough | **B** many | **C** more | **D** lots |
| 9 | **A** who | **B** which | **C** what | **D** whose |
| 10 | **A** jump | **B** climb | **C** flight | **D** trip |

## PET ❹ Writing

**Write a story of about 100 words beginning with this sentence:**

I was enjoying the flight when suddenly the aeroplane began to make strange noises...

T: GRADE 6

**5** Speaking: money

One of Amy Johnson's biggest problems was money. Like all pioneers and explorers she had to find financial support for her adventures. Today people who, for example, want to fly round the world in a balloon are helped by sponsors, usually big financial and industrial companies. Space projects and nuclear weapons are often sponsored/paid for by governments. Find some information or pictures which show how money can be spent in a useful way. Use the following areas to help you, and add others you may think of:

- education
- health and hospitals
- art and culture
- medical research.

 **INTERNET** PROJECT

Connect to the Internet and go to www.blackcat-cideb.com or www.cideb.it. Insert the title or part of the title of the book into our search engine. Open the page for *True Adventure Stories*. Click on the Internet project link. Go down the page until you find the title of this book and click on the relevant link for this project.

**A** The 1920/30s were the 'heroic age' of aviation. Find out about another famous female aviator, the American Amelia Earhart. Look for:

▶ her solo flight across the Atlantic in 1932

▶ her solo flight across the Pacific in 1935.

Organize your class into two groups. Each group can report on one of the above journeys.

**B** Amelia Earhart disappeared in 1937 during a round-the-world flight. Find out about:

▶ Amelia's journey and disappearance

▶ the theories about what happened to Amelia.

Each group can report on one of the above topics.

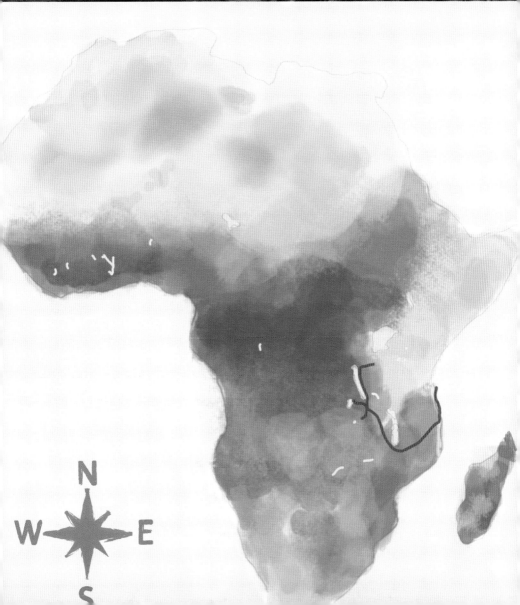

# David Livingstone's
# Last African Expedition

# Before you read

**1 Vocabulary**

**A   Match the words below (A-E) to their definitions (1-5).**

**A** expedition    **B** drug    **C** mud    **D** homesick    **E** journal

1   ☐   A chemical or organic substance used to make medicines.
2   ☐   A written record of daily activities.
3   ☐   Describes a strong feeling to return home.
4   ☐   A journey by land or sea with a special purpose.
5   ☐   Soft, wet earth.

**B   Find words in the text that match these definitions.**

1   A kind of bed for carrying a sick or injured person.

.........................................................................................................

2   A person whose job is to carry things, for example luggage.

.........................................................................................................

3   The place where a river begins.

.........................................................................................................

4   Describes people or places that are not friendly.

.........................................................................................................

**2 Listening**

**Listen to the beginning of Part One and answer the questions.**

1   What did Susi see on 10 November 1871?
2   What condition was Livingstone in when Stanley found him?
3   How did Stanley and Livingstone greet each other?
4   What did Livingstone hope to find on his expedition?

## PART ONE

# Always Searching but Never Finding

On the morning of 10 November 1871 one of David Livingstone's men, Susi, ran into his hut.

'Doctor! An Englishman has arrived!' he cried.

That day Livingstone was feeling very depressed [1]. Six years before, he had left England on an expedition to find the source of Africa's longest river, the Nile. Arriving in East Africa in 1866, the Scottish explorer had met with difficulty after difficulty. Now he was at Ujiji on Lake Tanganyika. He was ill, weak, very thin and hungry. And he thought the expedition had failed.

The white stranger was a journalist called Henry Morton Stanley, born in Wales in 1841, whose real name was John Rowlands. The editor of *The New York Herald* had sent him to Africa to find Dr Livingstone. In America and Europe people thought the famous missionary [2] and explorer was lost — or dead.

---

1.  **depressed** : 沮喪。

2.  **missionary** : 傳教士。

When Stanley saw Livingstone, he said, 'Dr Livingstone, I presume[1].' And they shook hands. It was one of the most famous meetings in the history of exploration.

Soon Livingstone was eating well and getting stronger. He felt much better. He was also more optimistic[2] about the expedition. Stanley's arrival had saved his life.

In April 1866, when Livingstone set out from the coast of East

1. **presume**：猜想。
2. **optimistic**：樂觀的。

Africa, he was already very famous for his African journeys. This search for the source of the Nile was his third expedition. He did not know that it was going to be his last adventure, and the most important one, too, because no white man had travelled along Livingstone's route before.

END

In June he reached Lake Nyasa with only a few African porters [1] and boys. Many had refused to go on. Livingstone continued his

1.　**porters** : 挑夫。

journey north. But at the end of October the rain began. It rained heavily every day. Everybody was very wet. And the food supply was getting low.

'We have neither sugar nor salt,' Livingstone wrote in his journal. 'I feel always hungry, and I dream about food all the time...'

Then in January 1867 two boys ran away with Livingstone's medicine box, which contained the drug quinine [1], necessary for fighting malaria, which Livingstone had. It was a disaster. Livingstone felt it was a 'sentence of death.'

But he did not stop. Wet and hungry, he went deeper into the unknown. When the expedition reached Lake Tanganyika, Livingstone was so ill with fever that he could not stand up. He was dangerously ill for some weeks. At the beginning of May he was ready to continue the journey again. But there was trouble among the people in the territory [2] where he was going, and he had to wait three months before going on to explore Lake Mweru. Then, in July, on the way to explore Lake Bangweulu to the southwest, Livingstone and his porters were nearly killed by some hostile [3] natives.

In December he set off from Lake Bangweulu to Ujiji to get food, drugs, and some news. The long journey in the rain gave him pneumonia [4]. He had a bad fever and could not walk, so his porters carried him on a stretcher [5]. In his journal for 8 March 1869 he says:

'I am very thin and I have no medicine... I hope to reach Ujiji...'

1. **quinine** : 奎寧。
2. **territory** : 地域。
3. **hostile** : 有敵意的。
4. **pneumonia** : 肺炎。
5. **stretcher** : 擔架。

Six days later he arrived in Ujiji and found that most of his supplies had gone: food, drugs and letters — all stolen. It was a terrible shock. But Livingstone had incredible courage and endurance. In July he started his journey of exploration again. He wanted to explore the Lualaba river in the dangerous territory around Nyangwe. He was hoping to find the source of the Nile there. This had become an obsession [1].

The rain did not stop. The local people were hostile. Porters ran away. There were always delays. Sometimes Livingstone had to walk through deep mud, or cross flooded rivers. Almost in despair, he went on for two years, always searching but never finding. On 1 January 1871 he was depressed, ill and homesick. At the end of January he heard that the men and supplies he needed had arrived in Ujiji at last. So he made his way there. On the way he witnessed a terrible massacre [2] of four hundred natives by some Arab slave traders.

When Livingstone arrived in Ujiji in July, he was very weak and feverish with malaria. He was as thin as a skeleton [3], but there was very little food left. And he had not found the source of the Nile.

And then Henry Morton Stanley arrived.

1. **obsession** : 困擾。
2. **massacre** : 大屠殺。

3. **skeleton** : 骷髏。

# The text and **beyond**

PET **1** Comprehension check

**Read sentences 1-10 and decide if each sentence is correct or incorrect. If it is correct, mark A. If it is incorrect, mark B.**

|  | | A | B |
|---|---|---|---|
| 1 | Stanley came to Africa to look for Dr Livingstone. | ☐ | ☐ |
| 2 | When Livingstone set out in April 1866, he already knew the route. | ☐ | ☐ |
| 3 | In January 1867, Livingstone was pessimistic about his survival because he had no food. | ☐ | ☐ |
| 4 | Illness and trouble among the African tribes delayed Livingstone at Lake Tanganyika. | ☐ | ☐ |
| 5 | The long trek from Lake Bangweulu to Ujiji took Livingstone fifteen months. | ☐ | ☐ |
| 6 | At Ujiji Livingstone found food, drugs, letters and news. | ☐ | ☐ |
| 7 | In July 1869, he almost discovered the source of the Nile. | ☐ | ☐ |
| 8 | He decided to go back to Ujiji when he heard that everything he needed had arrived there. | ☐ | ☐ |
| 9 | On the way to Ujiji he saw Arabs killing hundreds of native villagers. | ☐ | ☐ |
| 10 | Stanley arrived just at the right time. | ☐ | ☐ |

**2** David Livingstone

**Complete the text about Livingstone's life up to his last expedition with the sentences A-J.**

David Livingstone was born at Blantyre, near Glasgow, on 19 March 1813. In 1823, (**1**) ......., he left school and worked at the local cotton mill [1]. He worked 12-14 hours a day to save enough money to (**2**) ....... . There, he decided (**3**) ....... . In 1840, when he qualified in medicine, he went (**4**) ....... . Between 1840-56 he travelled (**5**) ....... Africa. He crossed the Kalahari desert, discovered Lake Ngami and crossed Africa (**6**) ....... .

1. **cotton mill** : 棉織廠。

(**7**) ....... in 1856 as a national hero, and in 1857 published (**8**) .......,
which was a bestseller. From 1858 to 1863 he explored the Zambesi
river valley, and (**9**) ....... . (**10**) ....... to give lectures on the Arab slave
trade in central Africa. In 1865 the Royal Geographical Society asked
him to return to Africa and search for the source of the Nile.

A   to Africa as a missionary.

B   go to medical school in Glasgow.

C   a book about his journeys.

D   In 1864 he travelled around Britain

E   when he was ten years old

F   discovered Lake Nyasa

G   from ocean to ocean

H   to become a doctor-missionary

I   He returned to England

J   thousands of miles across

**3** Writing

**Read this extract from Henry Stanley's autobiography（自傳） about
his famous meeting with Livingstone.**

> *I walked up to him and, taking off my helmet, I said, 'Dr Livingstone, I
> presume?'*
>
> *Smiling, he lifted his cap and answered, 'Yes.'*
>
> *Now that I was certain it was him, my face showed my happiness. I
> held out my hand and said, 'I thank God, Doctor, that I have found
> you.'*
>
> *He shook my hand warmly and replied, 'And I feel very happy that I
> am here to welcome you.'*

**Can you imagine how the conversation continued? Write a short
dialogue (about 60 words).**

# Before you read

**1** **Dangerous insects**

In Part Two you will read about two species of insects:

**tsetse flies:** African flies that suck the blood of people and cause serious diseases.

**red ants:** stinging insects that live in large groups.

**Which of the two insects would frighten you more? Why?**

**2** **Listening**

**PET**

Listen to the beginning of Part Two. If the sentence is correct, put a tick (✓) under A for YES. If it is not correct, put a tick under B for NO.

|     |                                                                                     | A | B |
|-----|-------------------------------------------------------------------------------------|---|---|
| 1   | In March 1872 Livingstone had enough porters for the expedition.                    | ☐ | ☐ |
| 2   | Stanley wanted Livingstone to continue searching for the source of the Nile.        | ☐ | ☐ |
| 3   | Livingstone's letter to *The New York Herald* was about the terrible massacres and the slave trade. | ☐ | ☐ |
| 4   | 14 March 1872 was the last time Stanley saw Livingstone.                             | ☐ | ☐ |
| 5   | The porters arrived five months after Stanley left.                                 | ☐ | ☐ |
| 6   | The expedition had to stop because food supplies were low.                          | ☐ | ☐ |
| 7   | Livingstone was too weak to walk.                                                    | ☐ | ☐ |
| 8   | Livingstone's men walked very slowly.                                               | ☐ | ☐ |

## PART TWO

# The Last Journey

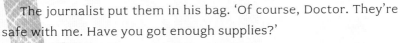

'Please take my letters and journals with you to Zanzibar, Mr Stanley,' said Livingstone.

The journalist put them in his bag. 'Of course, Doctor. They're safe with me. Have you got enough supplies?'

'Yes. Enough for four years!'

'When I arrive in Zanzibar, I'll find some porters and send them to you,' Stanley said. 'But won't you abandon your search for the source of the Nile? Come back with me.'

Livingstone smiled. 'No. If I go home now, I won't be well enough to come back and explore Central Africa. Anyway, after the terrible things I've seen here I want to do my best to stop the massacres and the slave trade. I've written a letter to your newspaper about it.'

So on 14 March 1872 Stanley said goodbye to Livingstone and left. He never saw the explorer again.

In August 1872 the porters arrived and the expedition set out from Tabora.

Soon conditions were bad again. Tsetse flies killed many transport animals. The sun made the ground too hot to walk on. Livingstone's health was not good. As the expedition moved towards Lake Bangweulu, the season of heavy rain began. He often got lost because the track was under water. Food supplies were low. At the end of the year the expedition had to stop; it was stuck in floods and mud.

Now Livingstone began to get weaker. He was so weak that he could not walk. His men had to carry him on their backs across rivers or through cold, wet fog. They walked only two and a half kilometres a day.

END

But Livingstone continued to write down his observations in his journal. He could still describe the wildlife, and red ants!

'The first ant came on my foot quietly,' he wrote. 'Then some began to bite between my toes. Then lots of big ones ran quickly over my foot and bit me so hard that the blood came.'

The expedition stopped at Lake Bangweulu. All around them was the flood water of the river Chambesi. Without canoes[1] they could not travel anywhere. Although he was very ill, Livingstone persuaded[2] the chief of a hostile tribe to give them canoes and food. So in heavy rain they crossed the Chambesi river.

Then they went south-west and reached the river Muanakazi on 12 April. Livingstone wrote this in his journal: 'I could only

1. **canoes** : 獨木舟。            2. **persuaded** : 說服。

walk a little way, and after two hours I lay down, exhausted.'

After a long rest he got up, walked along the muddy [1] track and then fell down. The men carried him on their backs again. When Livingstone tried to ride a donkey, he fell to the ground. His men knew that he was dying. They made a wooden stretcher with a canopy [2] against the sun, and carried him to Chitambo, a village near the southern end of Lake Bangweulu. They laid him inside a hut. There, on 1 May 1873, he died.

David Livingstone had reached the end of his long journey. He had not discovered the source of the Nile. But he had revealed the beauty and horrors of Central Africa to the world.

His African companions who had been with him on his last journey carried his body 2,415 kilometres back to the coast, a journey through hostile territory that took eight months. It was an act of homage [3] to the great explorer.

David Livingstone's body lies in Westminster Abbey.

1. **muddy**：泥濘的。
2. **canopy**：（遮陽）篷。
3. **homage**：尊崇。

## The text and **beyond**

**1 Comprehension check**

Look at the map of Livingstone's 1866-73 expedition and match sentences 1-10 with the places A-J.

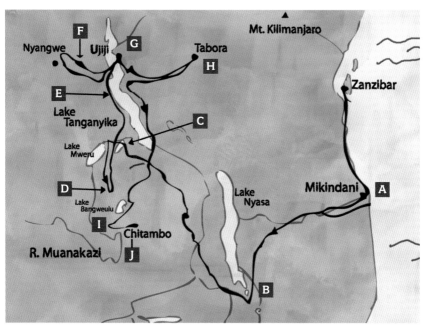

1 ☐ Livingstone decided to go back to Ujiji.
2 ☐ 10 November 1871. H.M. Stanley found Livingstone.
3 ☐ In June he reached Lake Nyasa.
4 ☐ 8 March 1869. 'I hope to reach Ujiji.'
5 ☐ Livingstone was dangerously ill with malaria.
6 ☐ August 1872. Livingstone set out on his last journey.
7 ☐ 1 May 1873. Livingstone died.
8 ☐ Livingstone set out in April 1866.
9 ☐ 12 April 1873. 'I could walk only a little way.'
10 ☐ 1871. Livingstone witnessed a massacre of 400 Africans.

**②  Summary**

**Put the sentences A-J in the correct order to make a summary of Part Two. Write 1-10 in the boxes.**

A ☐  After a stop at Lake Bangweulu they crossed the river Chambesi in canoes.

B ☐  But soon their animals died, the sun caused problems and Livingstone was in bad health.

C ☐  Then he tried to ride a donkey but fell off.

D ☐  Five months later, the explorer set off from Tabora on his last journey.

E ☐  In homage to Livingstone his men carried his body 2,415 kilometres back to the coast.

F ☐  By now Livingstone was so weak his men carried him on their backs.

G ☐  On 12 April 1872 Livingstone was so exhausted he had a long rest.

H ☐  Livingstone decided not to go back with Stanley when he left in March 1872.

I ☐  So his men made a stretcher and carried him to Chitambo, where he died.

J ☐  At the end of 1872 the expedition was stuck in floods and mud caused by heavy rain.

## 'He was so weak that he could not walk.'

In the example above the clause (從句) with **so... that** tells us the result of Livingstone's health condition. We can express the same idea using **too** and **enough**.

Note the different word order in the two examples below:

**too** + adjective (形容詞) + infinitive (不定式)

**not** + adjective (形容詞) + **enough** + infinitive (不定式)

*He was **too** weak **to** walk.*

*He was **not** strong **enough to** walk.*

Here is another example. Note how the first sentence changes:

*The ground was **so** hot **that** they could not walk on it.*

*The ground was **too** hot **to** walk on.*

We can also change a clause of result (結果從句) in this way:

*Livingstone was **so** ill **that** he died.*

*Livingstone died **because** he was so ill.*

This tells us **why** he died and is a clause of cause (原因從句) .

So:    *He could not walk **because** he was so weak.* (cause)

         *He was **so** weak **that** he could not walk.* (result)

**3** **Clauses of result**

**Change the following sentences using the words in brackets. There is an example at the beginning (0).**

0    The track was so muddy that they could not walk along it. *(too)*
     .The track was too muddy to walk along.......................................

1    They were hungry because the food supplies were so low. *(so... that)*
     ...........................................................................................

2    Livingstone was too weak to make a long trek. *(not... enough)*
     ...........................................................................................

3    They could not cross the river without canoes because it was very deep. *(too)*
     ...........................................................................................

4    Livingstone was too exhausted to ride a donkey. *(so... that)*
     ...........................................................................................

5    The rain was so heavy that they could not continue. *(because)*
     ...........................................................................................

6    'If I go home now, I won't be well enough to come back to Africa,' Livingstone told Stanley. *(too + ill)*
     ...........................................................................................

7    The weather was too wet to travel. *(not... enough)*
     ...........................................................................................

8    Livingstone fell off the donkey because he was so exhausted. *(so... that)*
     ...........................................................................................

**4 Discussion**

**Read the following questions and discuss your ideas with a partner.**

A   Explorers in Africa in the 19th century needed to have the following qualities:

- good health
- a very strong constitution
- a strong character
- determination
- ability for leadership
- the capacity to survive in heat, drought and tropical rain
- calmness when there was physical danger from hostile tribes or wild animals
- ability to endure many disappointments
- patience when things were slow or delayed
- desire to help free Africans from the slave trade
- a strong need to be famous.

1   Which of the above qualities do you think Livingstone had? Give reasons.

2   Which qualities do you admire most?

3   Do you think Livingstone's last expedition was a success or not? Give reasons.

B   Which of the above qualities do you think you have?

Which of the above qualities would you like to have?

# The Congo:
## *Heart of Darkness*

In 1902 the writer Joseph Conrad published a novel called *Heart of Darkness*. The story is based on his own experiences as a young man in 1890, when he travelled up the Congo River in a steamboat. He was one of the first Europeans to see the horrors of European imperialism [1] in the region. Conrad's story tells how a young captain, Marlow, sails up the Congo to rescue Kurtz, an agent for a European company trading in ivory [2]. Marlow finds a sick man who has become the native people's god. Using violence and murder, Kurtz forces them to bring the ivory that has made him rich. Around his hut are poles with human heads on them. Conrad wanted to say that the 'heart of darkness' is in the Europeans who exploited [3] the land and people of the Congo.

The Portuguese, thanks to the navigator Diogo Cao, were the first Europeans to reach the Congo River in 1482. They did not want to conquer or colonize [4] the Congo, but hoped to trade and introduce Christianity. Trade soon began, especially in slaves. Slavery was already part of the Congo culture. Local chiefs and Muslim merchants sold slaves to the Portuguese. Then in the 17th century the Dutch arrived, followed by the French and British. By the late 1600s European, Arab and African traders were transporting 15,000 slaves from the Congo.

1. **imperialism** : 帝國主義。
2. **ivory** : 象牙。
3. **exploited** : 剝削。
4. **colonize** : 殖民。

**Henry Morton Stanley** and the members of his expedition in 1874.

In 1874 the journalist Henry Morton Stanley led an expedition into central Africa, which had not been colonized or explored. After travelling 11,000 kilometres from the east coast, he reached the Congo River and travelled down it to the Atlantic Ocean, where he arrived in 1877. King Leopold of Belgium was one of the thousands of readers who followed the reports of Stanley's travels. At that time Belgium had no colonies and Leopold very much wanted to have a colonial empire. So, inviting Stanley to Belgium, he asked him to go back to the Congo as his agent. He told Stanley to get control of the rich trade in ivory. To do this, Stanley had to persuade local chiefs to sign documents which would give the land and people to Leopold. During the next five years Stanley signed 450 documents with Congo chiefs, who did not understand what they were signing in exchange for the cheap goods Stanley gave them.

In 1885 at a conference in Berlin, the European powers and the United States agreed to give Leopold the Congo River Basin as his personal property to do what he liked with. Over the next twenty-three years Leopold made a huge fortune from the trade in ivory and rubber. He formed the Force Publique, a private army of African soldiers led by European officers. By 1905 this army had grown to 16,000, which crushed any rebellion by the African workers. They burned villages, cut off the heads of chiefs and killed the women and children of men who refused to collect rubber and ivory. Going deep into the forest, they found and killed the rebels and cut off the right hand of every dead man to show how many they had killed. But the soldiers often cut off the hands of living people, including children, to increase the number. Finally, the workers went back to work and Leopold's profits rose higher and higher.

Then a shipping clerk named Edmund Morel discovered what was happening. He decided to expose the horrible atrocities [1] and his activities caused the British government to send a diplomat, Roger Casement, to investigate. He revealed evidence of mutilation [2], forced labour and murder. In 1904 Casement and Morel formed the Congo Reform Association, the first human rights movement of the 20th century, which spread to America. So public opinion turned against King Leopold, who had to give the Congo to the Belgian government in 1908. However, although the government stopped many abuses against the Congolese, it still controlled all the commercial interests and did not do much to help the people. Leopold's barbaric [3] rule was a disaster for the Congolese in every

1. **atrocities**：暴行。
2. **mutilation**：致殘。
3. **barbaric**：殘暴的。

way. 8-10 million people died from violence, forced labour and starvation [1] caused by his greed for power and profits.

In 1960 the Congo became an independent country. But an army leader, Joseph Mobutu, took power in 1965. Like Leopold, he stole the profits made from businesses in copper [2], diamond, oil and other resources. After the First Congo War, Mobutu was deposed in 1997 and the country was renamed the Democratic Republic of the Congo. But from 1998-2003 there was a disastrous Second Congo War, the worst conflict since World War Two.

Today fighting continues in the east of the country. The 'darkness' in the European heart has left a terrible shadow in the Congo.

1. **starvation**：餓死。
2. **copper**：紫銅。

**1 Comprehension check**

**For questions 1-10 choose the correct answer — A, B or C.**

**1** What is *Heart of Darkness* about?
- **A** ☐ Conrad's adventures in the Congo.
- **B** ☐ Kurtz's cruel exploitation of the Congo people.
- **C** ☐ Marlow's journey up the Congo River.

**2** Two hundred years after the Portuguese arrived
- **A** ☐ 15,000 slaves had been transported from the Congo.
- **B** ☐ the slave trade began.
- **C** ☐ the Portuguese and other nations were transporting thousands of slaves from the Congo.

**3** The reports of Stanley's expedition
- **A** ☐ inspired Leopold of Belgium to create a colonial empire.
- **B** ☐ did not interest many people.
- **C** ☐ were written by Stanley in 1877.

4  The Congo chiefs
   A ☐ received a lot of money for their land.
   B ☐ were persuaded to give away their land and people for
         nothing.
   C ☐ understood what they were doing.

5  From 1885-1908 Leopold
   A ☐ used the Congo as his private home.
   B ☐ made enormous profits from the ivory and rubber
         business.
   C ☐ ruled the Congo well.

6  How did some African workers react to Leopold's exploitation?
   A ☐ They refused to collect the rubber and ivory.
   B ☐ They hunted and killed the soldiers of the Force Publique.
   C ☐ They cut off their own hands so they could not work.

7  What was the result of Edmund Morel's activities?
   A ☐ the beating, mutilation and murder of the Congo workers.
   B ☐ the formation of the Congo Reform Association by the
         British government.
   C ☐ Roger Casement's investigation, which exposed the
         atrocities.

8  Leopold's crimes against the Congolese
   A ☐ stopped after 1908.
   B ☐ were a financial disaster for the Belgium government.
   C ☐ were supported by public opinion.

9  Joseph Mobutu
   A ☐ died in 1997.
   B ☐ ruled the Congo for over thirty years.
   C ☐ stole diamonds from the people.

10 Today the Democratic Republic of the Congo is
   A ☐ in the middle of a second civil war.
   B ☐ a peaceful country.
   C ☐ still suffering from the disasters of the past.

Taking **African slaves** on board a slave ship, an illustration from
*Cassell's History of England* (1930).

# The Slave Trade

During his journeys in Africa Livingstone saw Arab slave traders
capture [1] Africans to sell in slave markets on the coast. Angry and
disgusted [2], he was determined to tell the world about it and try to
stop it. As part of his missionary work he tried his best to end the
buying and selling of slaves.

At Lake Nyasa he saw Arab boats taking Africans across the lake.
Then the Africans were tied together and forced to march to the port
of Zanzibar, where they were sold and exported. With his own
money Livingstone paid for a boat to sail around the lake and guard
it like a police boat.

Slavery has existed everywhere for thousands of years. When the
Europeans first arrived in Africa, there was already a slave trade. The
Africans captured their enemies in war and sold them to Arab traders.
In Europe, the Greek and Roman civilizations had the greatest number

---

1.  **capture** : 捕獲。              2.  **disgusted** : 厭惡的 。

***An American slave market or The Auction Sale,***
an American painting of 1852.

of slaves in ancient history. The medieval [1] world had the fewest slaves. Then in the 1400s the slave trade began to grow again when Portuguese sailors first explored the coast of west Africa. They captured Africans, often with the help of other Africans, and sent them to Brazil to work on sugar plantations [2]. The Spanish also began to send slaves to the West Indies in the 1500s.

In the 16th and 17th centuries the slave trade grew even bigger as France, England and the Netherlands expanded their colonies. In 300 years Europeans transported about 10 million slaves from Africa to the New World. Nearly two million died on the way.

65% of slaves went to Brazil, Cuba and the West Indies to work on plantations. About 500,000 worked in the southern states of the USA on cotton and tobacco plantations. The journey across the Atlantic took months. The slaves were kept chained together in crowded, dirty conditions in the bottom of the ship, where many died of disease or cruel punishment. When the slaves were sold in the markets of the Caribbean and North and South America, the slave

1.  **medieval** : 中世紀的。　　　　2.  **plantations** : 種植園。

traders made enormous profits.

Slaves could not marry or own property and they had no civil rights. They had to work very long hours. In the southern USA a lot of slaves were treated badly, others well. But all of them were very important for the agriculture and economy of the south. In the north, free black people were also important in industry.

There was a rebellion of slaves under Spartacus in 73-71 BCE, but the biggest rebellion in history happened in Saint Dominique (now Haiti) in 1791. Half a million slaves rebelled against their French owners and took over the country. Then in the early 1800s people called abolitionists [1] began the fight to end slavery. In the English Parliament William Wilberforce (1759-1833) was the leader of the campaign [2] against the slave trade, which was abolished in 1807. But slavery did not finally end in the British colonies until 1839 and in the USA slavery was only abolished in 1865 at the end of the Civil War.

Today slavery is illegal in every country in the world, but it still continues in parts of Africa, Asia and South America.

1. **abolitionists** : 廢奴運動者。　　　2. **campaign** : 活動。

### ❶ Comprehension check
**Answer these questions.**

1　In East Africa in Livingstone's time where were the Africans taken and sold as slaves?

2　How did Livingstone try to stop the slave trade?

3　Which civilizations had the most and the fewest slaves in history?

4　Where did most of the slaves work between the 1400s and 1700s?

5　Why did nearly two million slaves die on the way to the New World?

6　What happened in Saint Dominique in 1791?

7　When did the abolitionists' campaign against slavery finally succeed?

#  **INTERNET** PROJECT

Find out more about the slaves' journey by ship. Organize your class into two groups. Each group can report on either A or B.

**A** The Middle Passage

- ▶ Why was it called 'the Middle Passage'?
- ▶ Approximately how long did it take?
- ▶ What nations were involved in the trade?
- ▶ What kind of goods were exchanged for the slaves?
- ▶ Approximately what number or percentage of slaves lost their lives during the passage?

Draw a triangle to illustrate the route of the slave ships. Label it with the information you have about the voyage from Europe and back.

**B** Conditions on a slave ship

- ▶ Approximately how many slaves did a ship carry?
- ▶ Describe the methods of chaining and the branding irons.
- ▶ What kind of food did the slaves eat?
- ▶ What were the causes of the high number of deaths?
- ▶ What happened to slaves who refused to eat or tried to commit suicide?
- ▶ Why did the ship's captain want to keep as many slaves alive as possible?

# INTERNET PROJECT

Connect to the Internet and go to www.blackcat-cideb.com or www.cideb.it . Insert the title or part of the title of the book into our search engine. Open the page for *True Adventure Stories*. Click on the Internet project link. Go down the page until you find the title of this book and click on the relevant link for this project.

Organize your class into four groups. Each group will work on one of the following:

▶ an adventurer/explorer who sailed round the world or survived the sea
▶ an explorer in the Antarctica or the South Pole
▶ a famous aviator
▶ an explorer in Africa.

Each group will prepare a report on one of the above subjects. The report will include information about the following:

▶ who the adventurer/explorer was
▶ their most famous adventure/expedition
▶ if their adventure/expedition was successful or not.

The report will also include some pictures from the Internet.

# EXIT TEST 1

## 1 Characters

Which of the people (A-L) in the stories do the questions 1-14 refer to? Write A-L in the boxes.

A   Mr Bell
B   Anne Robertson
C   Lyn Robertson
D   Amy Johnson
E   Henry Morton Stanley
F   Neil Robertson

G   Twenty-two of Shackleton's men
H   Dougal Robertson
I   Charles Lindbergh
J   Ernest Shackleton
K   David Livingstone
L   Lord Wakefield

**Who...**

1   ☐  worked as a journalist for an American newspaper?
2   ☐  survived the Antarctic winter on Elephant Island?
3   ☐  was a dreamer and had a gambler's spirit?
4   ☐  sat next to Shackleton at a dinner party?
5   ☐  was the first aviator to fly solo across the Atlantic?
6   ☐  could not forget his dangerous experience on the sea?
7   ☐  had a birthday on 4 July 1972?
8   ☐  was found alive in an African village?
9   ☐  decided to stay in the Bahamas with her boyfriend?
10  ☐  disappeared while flying over the Thames in 1941?
11  ☐   suggested to go round the world by boat?
12  ☐  said that an amateur pilot couldn't fly solo from England to Australia?
13  ☐  took off secretly from Croydon Airport?
14  ☐  was carried by his men across rivers?

## 2 Picture summary

Look at the pictures and the maps. Which of the four stories do they refer to? Write the title of the story under each picture.

........................................ ........................................ ........................................

........................................ ........................................ ........................................

........................................ ........................................ ........................................

........................................ ........................................ ........................................

........................................ ........................................ ........................................

........................................ ........................................ ........................................

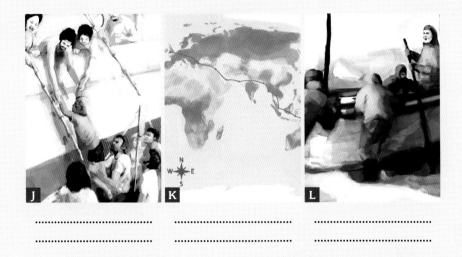

J ...................................
.................................

K ...................................
.................................

L ...................................
.................................

### 3 A graphic novel

Photocopy these two pages, cut out the pictures and stick them on paper. Think of words to put in the balloons when the characters are speaking or thinking. Do not use the words that were used in this book!

Then write at least a sentence under each picture to narrate what is happening.

### 4 Discussion

- Which story did you enjoy most?
- Why?
- Discuss your ideas with your partner.

# EXIT TEST 2

**1** **Answer the following questions.**

1 How many people were there on the *Lucette* when it sank?
2 Why did the Robertsons leave the raft?
3 Who rescued them?
4 What was Shackleton's great dream?
5 Where did the *James Caird* land on 11 May 1916?
6 When were the men on Elephant Island rescued?
7 What were Amy Johnson's qualifications?
8 Why did she want to fly solo to Australia?
9 What was the weather like when she flied across the Yomas Mountains?
10 Where was Livingstone when Henry Morton Stanley arrived?
11 When did the expedition set out from Tabora?
12 When and where did Livingstone die?

**2** **Are the following sentences true (T) or false (F)? Correct the false ones.**

|   |   | T | F |
|---|---|---|---|
| 1 | The *Lucette* sank because some whales attacked it. | ☐ | ☐ |
| 2 | On the raft the Robertsons had a compass, a survival kit, a knife and some food. | ☐ | ☐ |
| 3 | Dougal caught a shark with a spear. | ☐ | ☐ |
| 4 | There were two groups of men in Shackleton's expedition. | ☐ | ☐ |
| 5 | The *James Caird* landed on South Georgia island a week after it had left Elephant Island. | ☐ | ☐ |
| 6 | The twenty-two men on Elephant Island were rescued immediately. | ☐ | ☐ |
| 7 | Amy Johnson began her journey in 1929. | ☐ | ☐ |
| 8 | She landed at Aleppo on the third day. | ☐ | ☐ |
| 9 | She received £300 from the *Daily Mail* for her achievement. | ☐ | ☐ |
| 10 | Henry Morton Stanley was an American journalist. | ☐ | ☐ |
| 11 | Livingstone set out from the coast of East Africa in 1866. | ☐ | ☐ |
| 12 | Livingstone was carried to Chitambo on his men's backs. | ☐ | ☐ |

## *The Spirit of Adventure*

**Page 10 — activity 1**

1 T  2 F  3 F  4 T  5 T  6 F

## Lost at Sea

### PART ONE

**Page 12 — activity 1**

1 Falmouth  2 the Canary Islands
3 the Bahamas  4 Miami
5 Jamaica  6 Panama
7 the Galapagos Islands

**Page 12 — activity 2**

1 B  2 E  3 D  4 A  5 C  6 F

**Page 19 — activity 1**

A 8  B 3  C 5  D 9  E 1  F 6  G 10
H 4  I 2  J 7

**Page 20 — activity 2**

1 They left Falmouth in January 1971.
2 They began their voyage across the Atlantic.
3 Anne fell in love with a Canadian.
4 Dougal bought a dinghy.
5 Robin Williams asked them to take him to New Zealand.
6 They began their voyage across the Pacific.

**Page 20 — activity 3**

1 we were attacked  2 which / that Dougal bought  3 slept well  4 of survival were  5 don't we / not
6 Unless we / If we don't

## *The Enchanted Islands*

**Page 25 — activity 1**

1 T  2 T  3 F  4 T  5 F  6 F  7 F  8 T
9 T  10 F

### PART TWO

**Page 26 — activity 1**

*Open answers.*

**Page 26 — activity 2**

1 B  2 B  3 A  4 A  5 B

**Page 34 — activity 1**

1 The *Lucette* left the Galapagos Islands.
2 Killer whales attacked the boat and before it sank the Robertsons abandoned it.
3 Their clothes were in a bad condition, their skin was sore, they were tired, and their bodies ached.
4 Followed by two sharks Dougal had to swim to the dinghy and bring it back.

5 They moved from the raft to the dinghy.
6 They celebrated Lyn's birthday, but later a storm came.
7 Robin lost a turtle and Dougal hit him.
8 They were rescued by a Japanese ship.

## Page 35 — activity 2

1 B  2 B  3 A  4 A  5 A  6 B  7 A  8 B
9 B  10 A

## Page 35 — activity 3

1 On the third day a flying fish fell into the dinghy and they ate it for breakfast.
2 All day and night dorado swam against the raft and sometimes bit them.
3 On July 6th they caught a turtle with 100 eggs in it.
4 Dougal hit the shark with a paddle and it swam away.

## Page 36 — activity 4

A 4  B 5  C 2  D 6  E 1  F 3
A paddle  B ice cream  C (signal) flare  D spear  E knife  F dinghy

# The Impossible Journey
PART ONE

## Page 38 — activity 1

*Open answers.*

## Page 38 — activity 2

*Open answers.*

## Page 38 — activity 3

1 A  2 B  3 B  4 A  5 B  6 B  7 A  8 A
9 A  10 B

## Page 44 — activity 1

1 E  2 C  3 G  4 A  5 H  6 D  7 B  8 F

## Page 45 — activity 2

1 It was his great dream.
2 On January 19th 1915.
3 His ship was frozen into a huge piece of ice.
4 The ship was crushed and sank.
5 The ice was taking them away from the island/moving in the opposite direction.
6 Help from the sailors on whaling ships.
7 He rescued him from the water before the ice closed.
8 The wind changed, and Elephant Island was only 161kms away.
9 A big rock with cliffs, glaciers, and wind.
10 It was 1,400km across a very stormy sea.

## Page 45 — activity 3

*Open answers.*

## Page 46 — activity 4

1 B  2 B  3 A  4 A  5 B  6 A  7 A  8 B

## Page 47 — activity 5

*Open answers.*

## Page 48 — activity 1

A *Open answers.*
B 1 Perhaps the boat will capsize and sink.  2 Perhaps the men will fall overboard and drown.
3 Perhaps the men will lose hope and freeze to death.  4 Perhaps the men will die from exhaustion, or become weak and inefficient.
5 Perhaps the men will die of thirst.

**Page 48 — activity 2**

**1** T  **2** T  **3** F  **4** F  **5** T

## PART TWO

**Page 54 — activity 1**

**1** B  **2** A  **3** C  **4** B  **5** C  **6** D

**Page 55 — activity 2**

**1** are you  **2** Why don't you believe me? / Why not?  **3** you recognize  **4** are you doing  **5** crushed by the ice  **6** did you get  **7** walked across South Georgia to Stromness  **8** incredible to be  **9** They're waiting  **10** I'll send

**Page 56 — activity 3**

**1** C  **2** A  **3** C  **4** D  **5** A  **6** B  **7** B  **8** D  **9** C  **10** B

**Page 56 — activity 4**

**1** crawl  **2** gigantic  **3** survived  **4** thirsty  **5** tongues  **6** stream  **7** cold

**Page 57 — activity 5**

**1** journey  **2** deserts  **3** return  **4** ship  **5** Wall  **6** service  **7** home  **8** Venice  **9** jewels

**Page 57 — activity 6**

*Open answers.*

**Page 58 — activity 7**

*Open answers.*

## The Big Heat

**Page 63 — activity 1**

**1** B  **2** C  **3** C  **4** A  **5** A

## Amy Johnson, Woman Pioneer
### PART ONE

**Page 70 — activity 1**

1  The fact that Amy was a qualified mechanic.
2  He thought the idea of an amateur pilot flying to Australia was impossible.
3  Because in 1929 the public were losing interest in heroic solo flights.
4  She did not want publicity.
5  In the desert.
6  Because the local nomads sometimes attacked strangers.

**Page 70 — activity 2**

**A** 1  **B** 3  **C** 4  **D** 2

**Page 71 — activity 3**

**2** C  **3** D  **4** H  **5** A  **6** G  **7** E  **8** B  **9** A  **10** F

**Page 71 — activity 4**

**1** qualifications  **2** qualified  **3** engineer  **4** publicity  **5** dangerous/ courageous  **6** determination

## Bert Hinkler: the Lone Eagle

**Page 75 — activity 1**

**1** was fascinated  **2** how they flew  **3** to his back / tried to fly  **4** went to England / he wanted to make a career in aviation  **5** the Lone Eagle / a solitary man / liked to fly alone / any publicity  **6** shot down German planes **7** was abandoned  **8** he succeeded or created a world record  **9** the 1920s / from England to Riga / from Sydney to Bundaberg  **10** from New York to England via Jamaica, South America and West Africa / navigated with great accuracy / he used only a compass  **11** he died when his airplane crashed in Italy or his airplane crashed in Italy and he was killed

## PART TWO

### Page 76 — activity 1

**1** B  **2** A  **3** B  **4** B  **5** A  **6** A  **7** A  **8** B

### Page 76 — activity 2

**A** 966 km  **B** Karachi  **C** eight  **D** Yomas Mountains

### Page 81 — activity 1

**A** 7  **B** 4  **C** 1  **D** 5  **E** 10  **F** 2  **G** 8  **H** 3  **I** 11  **J** 6  **K** 9  **L** 12

### Page 82 — activity 2

**1** decided / had got  **2** had bought / went  **3** asked / had flown  **4** told / had already flown  **5** talked / had already helped  **6** arrived / had got  **7** could not beat / had never stopped  **8** thought / had crashed

### Page 83 — activity 3

**1** B  **2** A  **3** D  **4** B  **5** C  **6** D  **7** C  **8** A  **9** B  **10** A

### Page 83 — activity 4

*Open answers.*

### Page 84 — activity 5

*Open answers.*

# David Livingstone's Last African Expedition

## PART ONE

### Page 86 — activity 1

**A** 1 B  2 E  3 D  4 A  5 C
**B** 1 stretcher  2 porter  3 source  4 hostile

### Page 86 — activity 2

**1** an Englishman  **2** He was ill, thin, weak and hungry.  **3** They shook hands.  **4** The source of the Nile.

### Page 92 — activity 1

**1** A  **2** B  **3** B  **4** A  **5** A  **6** B  **7** B  **8** A  **9** A  **10** A

### Page 92 — activity 2

**1** E  **2** B  **3** H  **4** A  **5** J  **6** G  **7** I  **8** C  **9** F  **10** D

### Page 93 — activity 3

*Open answers.*

### Page 94 — activity 1

*Open answers.*

### Page 94 — activity 2

**1** B  **2** B  **3** A  **4** A  **5** A  **6** B  **7** A  **8** A

## PART TWO

### Page 99 — activity 1

**1** D  **2** G  **3** B  **4** E  **5** C  **6** H  **7** J  **8** A  **9** I  **10** F

### Page 100 — activity 2

**A** 6  **B** 3  **C** 8  **D** 2  **E** 10  **F** 5  **G** 7  **H** 1  **I** 9  **J** 4

### Page 101 — activity 3

1 The food supplies were so low that they were hungry.
2 Livingstone was not strong enough to make a long trek.
3 The river was too deep to cross without canoes.
4 Livingstone was so exhausted that he could not ride a donkey.
5 They could not continue because the rain was so heavy.
6 'If I go home now, I'll be too ill to come back to Africa,' Livingstone told Stanley.
7 The weather was not dry enough to travel.
8 Livingstone was so exhausted that he fell off the donkey.

### Page 102 — activity 4

*Open answers.*

## The Congo: Heart of Darkness

### Page 106 — activity 1

1 B  2 C  3 A  4 B  5 B  6 A  7 C  8 A  9 B  10 C

## The Slave Trade

### Page 110 — activity 1

1 To the coast at Zanzibar.
2 He paid for a boat to guard Lake Nyasa.
3 The Greek and Roman civilizations had the most slaves. The medieval world had the fewest.
4 Brazil, Cuba, and the West Indies.
5 The dirty, crowded conditions on the ships caused disease, and they were treated cruelly.
6 Half a million slaves rebelled and took it over.
7 In 1865 at the end of the American Civil War.

**Page 113 — activity 1**

1 E  2 G  3 J  4 A  5 I  6 H  7 C  8 K
9 B  10 D  11 F  12 L  13 D  14 K

**Page 114 — activity 2**

A *David Livingstone's Last African Expedition*  B *The Impossible Journey*
C *Lost at Sea*  D *Amy Johnson, Woman Pioneer*  E *Lost at Sea*
F *David Livingstone's Last African Expedition*  G *David Livingstone's Last African Expedition*
H *Amy Johnson, Woman Pioneer*
I *The Impossible Journey* J *Lost at Sea*
K *Amy Johnson, Woman Pioneer*
L *The Impossible Journey*

**Page 115 — activity 3**

*Open answers.*

**Page 115 — activity 4**

*Open answers.*

**Page 116 — activity 1**

1 Six people.
2 Because it couldn't hold them and their provisions.
3 Men from the *Tokamaru*, a Japanese ship.
4 To cross the Antarctic on foot by a route across the South Pole.
5 On South Georgia Island.
6 In September 1916.
7 She had a pilot's A licence and was a qualified mechanic.
8 Because it was a great challenge and she loved flying.
9 Stormy and Rainy
10 At Ujiji, on Lake Tanganyika.
11 In August 1872.
12 On 1 May 1873, in the village of Chitambo.

**Page 116 — activity 2**

1 T
2 F — They hadn't a compass.
3 F — He caught it with his hands.
4 T
5 F — It landed on South Georgia Island after 14 days.
6 F — They were rescued after 4 months.
7 F — In 1930.
8 T
9 F — She received £10,000.
10 F — He was from Wales.
11 T
12 F — He was carried on a stretcher.

NOTES

NOTES

NOTES

# NOTES

# Black Cat English Readers

# BLACK CAT ENGLISH CLUB
## Membership Application Form

**BLACK CAT ENGLISH CLUB** is for those who love English reading and seek for better English to share and learn with fun together.

**Benefits offered:**　　- *Membership Card*

- *Member badge, poster, bookmark*

- *Book discount coupon*

- *Black Cat English Reward Scheme*

- *English learning e-forum*

- *Surprise gift and more...*

Simply fill out the application form below and fax it back to 2565 1113.

**Join Now! It's FREE** exclusively for readers who have purchased *Black Cat English Readers* !

The book(or book set) that you have purchased: _____

English Name: _____ (Surname) _____ (Given Name)

Chinese Name: _____

Address: _____

Tel: _____ Fax: _____

Email: _____

　　　　　　　　　　　　　　　　(Login password for e-forum will be sent to this email address.)

Sex: ❏ Male　　❏ Female

Education Background: ❏ Primary 1-3　　　❏ Primary 4-6　　❏ Junior Secondary Education (F1-3)

　　　　　　　　❏ Senior Secondary Education (F4-5)　　❏ Matriculation

　　　　　　　　❏ College　　　　❏ University or above

Age: ❏ 6 - 9　　❏ 10 - 12　　❏ 13 - 15　　❏ 16 - 18　　❏ 19 - 24　　❏ 25 - 34

　　❏ 35 - 44　　❏ 45 - 54　　❏ 55 or above

Occupation: ❏ Student　　❏ Teacher　　❏ White Collar　　❏ Blue Collar

　　　　　❏ Professional　　❏ Manager　　❏ Business Owner　　❏ Housewife

　　　　　❏ Others (please specify: _____ )

As a member, what would you like **BLACK CAT ENGLISH CLUB** to offer:

　　❏ Member gathering/ party　　❏ English class with native teacher　　❏ English competition

　　❏ Newsletter　　　　　　❏ Online sharing　　　　　　❏ Book fair

　　❏ Book discount　　　　　❏ Others (please specify: _____ )

Other suggestions to **BLACK CAT ENGLISH CLUB**:

_____

Please sign here: _____

(Date: _____ )